FOR DANIEL AND BEN

Contents

Preface

Certain times have been pivotal in the history of life on Earth. We live in such a time today, as human society appropriates 40 percent of the plant growth that takes place each year. This massive impact threatens our evolutionary heritage—the rich variety of animal and plant species on Earth that is called "biodiversity." Today there are approximately four hundred thousand flowering plant species, but each year their natural habitats shrink and are fragmented by human encroachment; as a result, a quarter of all plant species are presently at risk of extinction. Preventing these extinctions is an urgent task, but so is understanding what we might lose. Why *are* there so many plant species?

How did evolution furnish the Earth with such variety? Could evolution replenish what we are now destroying, or might we have so mishandled nature that she will devour her own creations, leaving a world overpowered and impoverished by the survival of only the fittest? Evolution *is* change. Every question about the rise of diversity or its demise is, at a fundamental level, an evolutionary question.

Certain places are at the fulcrum of change. According to biblical fable the Garden of Eden was such a place, where Adam and Eve, the ancestors of the human race, ate the forbidden fruit of the tree of knowledge and the whole species was exiled from the Garden in punishment. But, we still live in a garden—there is after all no life without plants to feed us and our livestock. And a new tree of knowledge is growing, tended by scientists intent on tracing the genealogy that links all living species. Ponder the astounding fact that not only are all members of the human species close kin, but that we are also kin to our pet goldfish and dogs, to the tomatoes ripening on the windowsill, and to every plant and insect in the garden.

All life today probably originated from a single evolutionary event that took place about 3.5 billion years ago. We know all life shares a common ancestor because there is no other rational explanation for why all living things have the same basic biological processes. In particular, all life is based on the DNA molecule that stores the genetic code and transmits it from one generation to the next. If eating from the biblical tree of knowledge led to our expulsion from the Garden of Eden, then tasting the fruit of the tree of evolutionary knowledge is a ticket of readmission. It makes us realize that we are, and always have been, a part of that first Garden. In truth there was never any expulsion, only ignorance of where we truly belong in the scheme of nature.

This book germinated and grew with the speed of a weed, but finding the right publisher was an endurance test requiring the tenacity of a hardy perennial. Happily, Christie Henry at the University of Chicago Press had the perspicacity to see in *Demons in Eden* what others would or could not, and so here it is. I am grateful to the many people who read some or all of the manuscript and to those who offered me help and encouragement along the way: Willam Allen, Peter Ashton, Spencer Barrett, Gerald Bristow, Steve Bullen, Mark Chase, Jonathan Cobb, Rodolfo Dirzo, Mike Dodd, Roland Ennos, Javier Francisco-Ortega, Miguel Franco, John Geiger, David Gowing, Peter Grubb, Steve Hubbell, Johnny Johnston, Richard Law, David Leach, Justine Leach, Don Levin, Hong Liu, Diethart Matthies, Kay McCauley, Eric Menges, Adrian Newton, Rissa de la Paz, Caroline Pond, Paul Ramsay, Jed Redwine, Irene Ridge, Steven Rose, Arnoldo Santos-Guerra, Adrian Silvertown, Pam Solomon, Carly Stevens, Cathy Trodd, Ian Sherman, Colin Walker, Cam Webb, and Mark Williamson. Responsibility for any errors and omissions is entirely mine.

I also wish to thank my research collaborators who made unwitting contributions to this book through some of the research described here. In particular, Mike Dodd, David Gowing, and Kevin McConway at the Open University and Takashi Kohyama, Nikki Kachi, and Etsushi Kato in Japan. I am grateful to the Open University, the Natural Environment Research Council, the Royal Society, the British Council, and the Japanese Society for the Promotion of Science for research and travel funding at various times. Above all, I thank my infallible guide in all things, my wife Rissa de la Paz, to whom I owe everything.

I

An Evolving Eden

At a spot along the banks of the River Thames, in the southwestern suburbs of London, there "[s]its enthroned in vegetable pride, Imperial Kew by Thames' glittering side." So wrote Erasmus Darwin, Charles's grandfather, in his book-length poem titled *The Botanic Garden*. The imperial days of Kew Gardens are gone, but in its three hundred acres more than twenty thousand plant species are found in cultivation. It is a good place to sample plant diversity—to wander among the plants and wonder just why there are so many.

Stroll across Kew Green, an oval expanse of grass where cricket is played in summer, toward the imposing iron gates of the garden erected by Sir William Hooker in 1845. Passing through the turnstile, one encounters verdant English lawns interspersed by groves of oaks, bamboos, pines, and other trees in all their variety; spectacular beds of flowers that change kaleidoscopically with the seasons; alpine rockeries that are more richly stocked than the Alps themselves; and even, tucked away in a corner, a little garden displaying a palette of a dozen species of lawn grass, each a subtly different, textured shade of green from its neighbor. And there is much more. The famous Victorian Palm House contains a tropical forest and in its basement are illuminated aquariums stocked with marine algae. The Temperate House has plants from all temperate regions of the globe, but the huge and modern Princess of Wales Conservatory outdoes them all.

Enter by the door at the southern end where pelargoniums and jade plants from South Africa grow just inside. These are familiar as houseplants and the temperature is cool and pleasant, but be ready for more exotic sights. A step farther, through more doors, and you are in the dry tropics among desert plants from the Americas and southern Africa. These are habitats where plants struggle with a physical environment that is blisteringly hot by day, nearly freezing at night, and parched most of the year, and where any unguarded green substance is a moist lunch for a herbivore. Evolutionary adaptations to the twin rigors of an unforgiving physical environment and hostile herbivores are evident in every plant of the desert.

Nestled in a stony bed are the clublike leaves of "living stones" (*Lithops* species) from the Namib Desert. Most of the plant is buried below ground level, and all that shows are the flattened tops of a pair of clubs, separated by a cleft. The crown of each club is brown and mottled, providing camouflage against a stony background. In fact the mottling hides a window that admits light into the interior of the plant. In the dry season the leaves shrivel, to be replaced by new ones when the rains come. Hiding from herbivores and avoiding the heat of the sun are all very well, but it doesn't do to hide from pollinators. In their season, bright flowers issue from the cleft in the living stones to claim their fifteen minutes of conspicuous fame.

Continue along the path that rises and turns a corner and then, suddenly, you are in a theater of desert plants. The cast is international, but all are actors in the same evolutionary play whose theme is convergent evolution. Center stage is a gigantic specimen of the century plant, *Agave americana*, from Mexico. It is seated, like a many-tentacled beast, on a platform of rock, its lower leaves arched downward as if drooping under their own weight. Each leaf is like a massive, grey-green spearhead, nearly two meters long and tipped with a needle-sharp spine. In the center of the plant, pointing straight at the sky like a challenge, are three waxy new leaves with freshly hardened spines. This is an awesome plant and only a macho Mexican armed with a machete would tangle with it.

Framing the century plant, its naked limbs tottering to a height of five meters, is a desert tree from Africa, *Aloe barberae.* Its limbs are striated with a pattern of light brown sinews and each trunk terminates in a rosette of succulent, dark green leaves. This aloe-on-a-stick is an eccentric plant. Some of its more terrestrial relatives are remarkably like *Agave*, with a rosette of stiff spearlike leaves and no trunk. Aloes in Africa and agaves in America belong to quite different plant families, but they occupy similar habitats on their respective continents and have independently evolved similar life forms. They are an example of convergent evolution. When environmental conditions are similar in different parts of the world, evolution has often fashioned similar looking organisms in each of them, but using different, local starting materials. Thus, in an Old World desert the spiny rosette plants belong to the aloe family, while in deserts of the New World the same growth form has evolved in the agave family. Outward similarities conceal different evolutionary origins.

In the footlights of the desert stage sits a cluster of barrel cacti from Mexico, the size of overlarge basketballs, that repeat the theme of convergent evolution with another growth form. The barrel shape of the cactus has pleats that allow the plant to expand when it takes up water. An unprotected barrel full of water has as much chance of survival in the desert as a snowball in hell. Barrel cacti are guarded by a thicket of ferocious yellow spines that protect against herbivores, but unfortunately not against people who wish to collect such spectacular plants. The largest specimens at Kew were confiscated from an illegal shipment that was intercepted by UK Customs and Excise.

The cacti of the American deserts have their parallel in Africa in plants belonging to the family *Euphorbiaceae.* Like barrel cacti and others, many African desert euphorbs have spiny, fluted stems. Indeed, these plants look so much like cacti that they are often mistaken for them, though true cacti are native only in the Americas. The separate evolutionary origins of the cactuslike growth forms in Africa and America are disguised by convergent evolution. Features of their flowers that have not been shaped by convergent evolution tell them apart.

In the wings of the desert stage hangs a third example of evolutionary convergence between the desert flora of Africa and its equivalent in the New World. An African climbing succulent belonging to the milkweed family scrambles up a training wire, but looks no more than a thick wire itself: it is all stem and no leaves. At a height of about two meters, this African plant flops over into the spinous canopy of a South American tree, from which its jointed stems hang vertically downward as if tired of resisting gravity. Perhaps the supporting tree has a sense of *déjà vu* because there is a genus of cacti called *Rhipsalis* that trails their jointed, leafless, and almost spineless stems in an identical manner in South America.

Paradoxical though it sounds, convergence is one way in which evolution generates diversity. By recreating the same type of plant in different geographical regions, evolution has produced not just one group of cactuslike plants but two: one in the New World and another in the Old. Rosette plants and succulent, leafless climbers have likewise been duplicated. Thus, geographic barriers between regions have caused whole floras to evolve in isolation from one another, enriching the global flora. How local biodiversity can suffer when these barriers are breached by plants being taken from one place to another we shall see later.

Leave the dry tropical zone by a flight of steps that gives a parting view of the desert theater from backstage, and pass through a glass door. You are enveloped by a hot, clammy fog. There is a hiss overhead from the vaporizers that keep the humidity of this part of the conservatory at a level more to the liking of its inhabitants. Through the mist you decipher a sign that reads "ZONE 1: Moist Tropics." This is another environment with another whole set of species, collected together here from all over the Old and New World tropics. In nature, moist tropical environments contain more plant species than any other on Earth. Just three of the most spectacular species to be found at Kew illustrate the role of insects in the evolution of plant diversity. There is a good view of the first one over the balcony just ahead. Look down from this vantage point into a pool of clear water where the giant water lily *Victoria amazonica* reigns. Each lily pad is like a giant's frying pan, up to two and one-half meters across, with an upturned rim that helps it float. The stalk and lower surface of the leaf are armed with large spines. This plant's enemies are aquatic and attack from beneath, while its pollinators are aerial creatures.

Victoria amazonica has flowers that befit a giant. They are about a foot across and also protected by spines when in bud. Sir Ghillean Prance, a former director of Kew, discovered how the giant lily is pollinated in the wild. Its secret attraction for pollinators is in the heart of the flower where there is a hollow chamber with the stigma (female part) perched in its base and a ring around the interior of the chamber roof stocked with starch and sugar. Flower buds develop below the surface and when ready to open find their way upward toward a window of light between the lily pads. In the daylight hours before its nocturnal debut the flower stalk pushes the bud, still tightly closed, clear of the water surface where it waits for sunset. As the light fades, the flowers slowly open and release a sweet fragrance: an entire lake of unfolding Amazon lilies is a breathtaking sight. Enticed by the perfume, and guided by white petals that shine like beacons in the moonlight, scarab beetles mob the flowers and crawl through a tunnel into the inner chamber where they feed on the starch and sugar to be found there. At about midnight the flowers close, trapping the feasting beetles inside, and soon afterward the petals begin a blushing transformation of color. By midafternoon of the following day, the flowers that closed with white petals have reopened in velvet-purple, but the beetles are still held prisoner because the tunnel from the central chamber of each flower remains blocked. At sunset the tunnel is unblocked and as they scramble to escape, the beetles, sticky from their honeyed meal, become covered in pollen from anthers that have now opened in the lining of the escape tunnel. As they fly off to another night of sweet imprisonment, the beetles carry with them pollen that they will deposit on the stigma in the chambered heart of another flower.

One insect's meat is another insect's poison. Thus, a diversity of tastes among pollinating insects can engender the evolution of a diversity of flowers among plants. While sweet perfume attracts beetles in search of nectar, the rank smell of rotting flesh brings flies. What *is* that smell? It seems to be coming from another pool, upstream of the lily pond. Cascading from a tree is a climber with large heart-shaped leaves that some joker seems to have hung with surreal saxophones made of flimsy white cloth spattered with dried blood.

Few artists can match evolution's flare for the bizarre and those who try usually plagiarize nature. But this is no artefact. These are pelican flowers, belonging to a vine from South America. The face of the flower is heart-shaped in outline, about half a meter from top to bottom, with an eighty-centimeter-long tassel hanging from its lower lip. Just above the center of the flower is an opening stained a deep magenta, and all around it the same bloody hue flecks its creamy fabric. This is a lure for flies. A plant mimic of a murder scene, honed by evolution into the quintessence of a fly's wildest desires, reproduced in full smell-o-rama. The flower is like a flared horn and its magenta opening is the entrance to a tube that bulges and then goes through a U-shaped bend, like a saxophone. Where the mouthpiece of the saxophone would be, the tube narrows to a stalk by which it hangs from the vine.

One final plant to see before leaving the conservatory is the titan arum, which has one of the biggest, smelliest flowers on Earth. Perhaps fortunately the titan arum isn't actually flowering right now because otherwise there would be a long line of visitors waiting to see (and smell) this rare event, patiently filing past its stinking, two-meter-tall flower spike, guarded by a specially assigned policeman from the Kew Gardens Constabulary wearing a gas mask. The leaf of this plant (not leaves, for it has but one) is not noisome but is just as spectacular as its inflorescence. There are three of these plants in the conservatory at the moment, each in a massive plant pot. Out of each pot sprouts a stout leaf stalk that would respectably serve a medium-sized tree for a trunk. The stalks lean at about thirty degrees from vertical and are perfectly smooth, dark green in color with an attractive pattern of lighter green splotches. At a height of three meters or so the stalk branches into leaflet-bearing ribs, each rib dividing into two, then four, then eight. The result is a shallow-domed canopy like an upturned saucer covered in a single layer of foliage. This leaf feeds an underground tuber that will grow to thirty kilograms in weight before it is ready to flower. To what extreme lengths some plants are driven by the evolutionary imperative to reproduce!

The giant water lily, the pelican flower, and the titan arum are spectacular examples of how pollinators can drive the evolution of flowers in weird and wonderful directions. Though these plants have exceptionally large flowers, the significance of insects in their evolution is by no means exceptional. Interactions with pollinating insects have been a major diversifying force in the evolution of the flowering plants and almost certainly account for why this group as a whole is so numerous while the conifers, which are wind pollinated, number fewer than a thousand living species.

A visit to the Princess of Wales Conservatory gives a tantalizing taste of the wonderful diversity of the plant world. Though unusual species grab the attention, there isn't any plant, not even the garden-variety daisy or dandelion that is heedlessly trodden underfoot, that doesn't have an evolutionary story to tell. Some of them will be heard later in this book. All are part of a single evolutionary drama—the story of how the green branches of the evolutionary tree sprouted and grew into the diversity of plant species we enjoy today.

Botanical gardens have an important role in researching and preserving botanical diversity, nowhere more so than Kew. But Kew also has a unique place in the history of evolutionary thought owing to the part it played in the development of Charles Darwin's ideas on evolution. Darwin lived and wrote *On the Origin of Species* not twenty-five miles from Kew at Downe, in Kent. The link between the two was Joseph Hooker, who in the mid-nineteenth century assisted and eventually succeeded his father as director at Kew. He was a confidant of Charles Darwin and a key figure in the development of Darwin's theory.

Like Darwin, who had undertaken a round-the-world voyage in his early twenties, Joseph Hooker at a similar age was assistant surgeon and botanist on a four-year expedition to the Southern Hemisphere. Soon after Hooker's return, Darwin seems to have recognized him as a kindred spirit, addressing him quite early in their correspondence as a "co-circum-wanderer and fellow labourer." In 1844 Joseph Hooker was the first scientist to whom Darwin confessed his conviction that species were not separately created and immutable, but rather had evolved one from another and shared a common descent. Darwin wrote in a letter to Hooker that, after much research, "[a]t last gleams of light have come, and I am almost convinced (quite contrary to [the] opinion I started with) that species are not (it is like confessing a murder) immutable."

Like "confessing a murder"? These words seem strange to modern ears, but indicate how guilty Darwin felt that he was straying far from the scientific and theological orthodoxy of his day. The prevailing view was that species were individually created by God and that adaptation, the marvelous way in which plants and animals are suited to their various roles in nature, was irrefutable evidence for the existence and goodness of the Creator. To challenge the immutability of species was to challenge the very foundations of Victorian religion and society. It was to challenge God Himself. Darwin's grandfather Erasmus had been publicly ridiculed for his evolutionist views; his father, Robert, had kept similar ideas private to avoid trouble. Now Charles, fearful of the reaction he would provoke from the establishment, found himself forced by the scientific evidence to reluctantly tread the same path of heresy. He did so very cautiously. In fact Darwin's theory of natural selection was first outlined in his notebook at the end of 1838, but such was his caution that he would wait five more years before even hinting at it in the letter to Hooker.

Fifteen more years elapsed between his confessional letter to Hooker and the eventual publication of Darwin's theory in *On the Origin of Species*. Darwin spent much of those fifteen years painstakingly and secretly accumulating an unassailable mountain of evidence, and he frequently turned to Hooker at Kew for botanical and other facts and opinions to support his case. In 1859 in the introduction to *The Origin* Darwin acknowledged "the generous assistance which I have received from very many naturalists," but only Hooker was acknowledged by name.

The idea that species change, though heretical in mid-Victorian England, was by no means new. Charles's own grandfather Erasmus had propounded it in verse in his 1803 poem *The Temple of Nature.* Lamarck, in revolutionary France, was an evolutionist, but English naturalists including Darwin rejected his ideas on adaptation as absurd. What made Charles Darwin's theory different, and what gave it enduring value, as he claimed in his letter to Joseph Hooker in 1844, was that "I think I have found out (here's presumption!) the simple way by which species become exquisitely adapted to various ends." Natural selection, the *mechanism* of evolution, is the quintessence of Darwinism and explains how adaptation arises without God.

The Darwinian mechanism requires three ingredients for it to work. First, there must be *variation* among the individuals of a species. It is a biological fact that even closely related individuals differ from one another in all sorts of subtle and not-so-subtle ways. Even identical twins, so alike in many ways, do not have the same fingerprints. Before *The Origin*, variation within species was regarded as having little significance, but to Darwin it was the essential raw material of evolution. The second ingredient in Darwin's mechanism is *heredity.* Some of the variation observed in nature is inherited, being passed from parents to offspring. Eye color in humans is an example. Finally, *selection* must favor some inherited variants over others. Flower color is often inherited. In wild radish, for example, pollinating insects prefer to visit yellow-flowered plants rather than white ones of the same species. Consequently, yellow-flowered plants produce more seeds than white ones and thus have a selective advantage that works to displace them.

Natural selection is driven by the pressure of population increase that produces a struggle for existence. In Charles's words:

> A struggle for existence inevitably follows from the high rate at which all organic beings tend to increase. . . . There is no exception to the rule that every organic being naturally increases at so high a rate, that if not destroyed, the earth would soon be covered by the progeny of a single pair.

Or as Charles's grandfather, Erasmus, put it in verse in *The Temple of Nature:*

> Each pregnant Oak ten thousand acorns forms
> Profusely scatter'd by autumnal storms . . .
> . . . All these, increasing by successive birth,
> Would each o'erpeople ocean, air, and earth.

The result, says Erasmus, is a war among plants:

> Yes! Smiling Flora drives her armèd car
> Through the thick ranks of vegetable war;
> Herb, shrub, and tree, with strong emotions rise
> For light and air, and battle in the skies;
> Whose roots diverging with opposing toil
> Contend below for moisture and for soil;
> Round the tall Elm the flattering Ivies bend,
> And strangle, as they clasp, their struggling friend.

The grandson of the evolutionary poet defined the struggle for existence in a broader sense, with an emphasis on the victors as being those that leave the most offspring: "I use the term Struggle for Existence in a large and metaphorical sense, including dependence of one being on another, and including (which is more important) not only the life of the individual, but *success in leaving progeny*" (emphasis added).

Progeny are the prize of evolutionary success and multiply the rewards of natural selection with each passing generation. A successful variant can spread with amazing rapidity. Charles adduced the evidence from a higher botanical authority: "Linnaeus has calculated that if an annual plant produced only two seeds—and there is no plant so unproductive as this—and their seedlings next year produced two, and so on, then in twenty years there would be a million plants." But Darwin had also seen with his own eyes in Argentina how quickly successful reproduction could alter a landscape: "Cases could be given of introduced plants which have become common throughout whole islands in a period of less than ten years. Several of the plants now most numerous over the wide plains of La Plata, clothing square leagues of surface almost to the exclusion of all other plants, have been introduced from Europe."

There is a paradox contained in the Darwinian argument, and it is this. Natural selection favors those individuals that leave the most offspring. The descendants of these individuals inherit the advantages of their parents and continue to multiply, while the progeny of others become fewer and fewer until they are gone. Does this sound like a mechanism that would generate the diversity of plants seen in Kew Gardens or elsewhere? Hardly! Rather, it sounds like a mechanism designed to favor only a creature with superlative powers of reproduction: a Darwinian demon. Where is there room for anything else in the struggle for existence?

There is a potential Darwinian demon hiding in every species because all populations are capable of increasing geometrically if unchecked. When he visited Argentina on *HMS Beagle* Darwin saw what plants out of control could do in the European species that had romped across "the wide plains of La Plata."

Plenty more Darwinian demons from Europe and Asia have now been unleashed upon the Americas. One of the most widespread of these is purple loosestrife (*Lythrum salicaria*), which arrived on the eastern seaboard of North America in the early 1800s. In Europe, this attractive plant with masses of pink-purple flowers is found by the margins of lakes and streams and poses no particular problem to anybody. Native populations of purple loosestrife sometimes reach densities that exclude other species, but this dominance is local and temporary. In North America, by contrast, the plant forms solid stands that seem to be self-replacing and permanent, and it has conquered millions of hectares of wetlands and displaced native plant species, reducing some to the brink of extinction.

What turns a plant like purple loosestrife that is well behaved at home into such a demon when it ventures abroad? It has prodigious fecundity, good seed dispersal, and vigorous growth, all of which may predispose purple loosestrife to demonhood, but they do not preordain it because the plant is no demon in its native Europe. Why is this potential only fully realized in new territory? Why are this and other alien plants such demons? Florida, which has worse problems with alien flora and fauna than many places, holds some of the answers, as we shall see in chapter 8.

A rapid growth rate is a demonic trait that makes some plant species a particular threat, but it is not sufficient on its own to turn a plant into a demon. An example all too familiar to British gardeners is the Leyland cypress. This tree is a hybrid between two American conifers: Monterey cypress, *Cupressus macrocarpa*, from California and the Alaskan cedar, *Chamaecyparis nootkatensis*. The American species encountered one another on Welsh soil where a hybrid seedling was first found at Leighton Hall in 1888 and was named after its proprietor, C. J. Leyland. The scientific name of the plant, *Cupressocyparis leylandii*, betrays its hybrid origin and commemorates its discoverer.

C. leylandii turned out to have a phenomenal growth rate, able to put on as much as four feet (one and one-third meters) of height in a year and to easily reach heights of one hundred feet (30 meters). Indeed, it is still not certain what will be the maximum height this new species can reach. With its rapid growth and dense, evergreen foliage, Leyland cypress can produce an almost instant garden hedge, but frequent trimming is needed to prevent the instant hedge turning into a permanent nightmare for neighbors, plant, and human alike. So common is this problem in Britain that in 1999 the environment minister told a reporter from the *Guardian* newspaper that "[i]n two years as minister of the environment the subject of *leylandii* has dominated my postbag. Forget climate change, sea level rise and all the other issues, desperate and oppressed people write to the government for help because they have nowhere else to turn." An Act of Parliament was passed in 2003 to deal with the problem of disputes between neighbors over the size of hedges made from *C. leylandii*.

Leyland cypress can appear demonic to the neighbors, but it is not a real Darwinian demon because it cannot fend for itself in the wild. It is only a problem where it has been thoughtlessly planted. However, could this tree be a warning of what might happen in future with other new plants? *Leylandii* is a hybrid between two other species and appears to owe its amazing growth rate to a combination of the genes of its parent species. Genetic engineering is now used to transfer genes between species in order to produce plants with novel traits. Could genetic engineering unleash a genuine, fully qualified Darwinian demon? This question is explored in chapter 9.

C. Leylandii fails to qualify as a fully fledged demon because it needs help in leaving progeny. Darwin noted in *The Origin* that this was the case for a huge number of garden plants. But there are exceptions, and these are the real demons: plants such as the kudzu vine in Florida that was introduced as an ornamental from Japan and now runs rampant across the southern United States (see chapter 8), or *Rhododendron ponticum* brought to Britain from Turkey. This shrub now rampages through certain British woodlands, extinguishing the native flora with its dense shade and acid leaf litter.

The big question is simply the converse of the alien plant riddle: What holds in check the demonic tendencies of most species? Darwin recognized in *The Origin* that he must address this crucial question: "What checks the natural tendency of each species to increase in number is most obscure. Look at the most vigorous species; by as much as it swarms in numbers, by so much will its tendency to increase be still further increased." Or, in other words, abundant species should become ever-more numerous. Why don't they?

The two sides to this problem are fundamental to the evolution of diversity and to its survival. To understand the evolution of diversity we need to solve the mystery of how, on occasion, evolution enables plants to escape their ancestral limits and to form a new branch of the evolutionary tree. If it is to spread, a new species must have its day as a demon, however local or temporary its dominance. Conversely, to understand why diversity is not eroded by every new demon, we need to solve a second mystery: What puts the brakes on demons and prevents their success causing the destruction of diversity? These are the two themes of this book. They are the *how* and the *why* of plant biodiversity—how it evolves and why it persists.

These are abstract ideas, but the power of the theory of evolution and of our conceptual understanding of ecology must be measured against their ability to explain the way the world is. We shall therefore explore how biodiversity evolves and why it persists by visiting real places to look at real plants in search of answers to these questions. In most cases, I retrace the steps I have taken during my own research into these questions over the last thirty years. This is a personal odyssey as well as a scientific one.

Our first stop is the laboratories in Kew Gardens because this is one of the places where the new tree of knowledge—the evolutionary tree of the plants—is being cultivated. There is no better place to grasp the grand picture of plant evolutionary history as it is now understood. We then leave England for the Canary Islands, a workshop of evolution located one hundred kilometers from the Atlantic shores of North Africa. Evolution seems to spew new plant species on these islands like volcanos spew fire and the plants there can show us how plant diversity evolves.

Sparing no expense and unconstrained by the logistics of real travel, I next take you to Japan where we will climb Mount Shimagare, an extinct volcano whose slopes are inhabited by active Darwinian demons. This is where we discover what puts the brakes on demons' power to destroy. This knowledge is the first key to the mystery of why diversity persists. From the demon mountain, we travel in chapter 5 to the tropics, a realm of unparalleled plant diversity, where we investigate some of the latest ideas about how the huge numbers of tree species found in tropical forest are able to live with one another. Then, in chapter 6, it is back to England to find out whether the same ideas can explain the persistence of plant diversity in species-rich grasslands and meadows. Can the keys to plant diversity discovered in the tropics unlock the mystery of meadow plant diversity in a temperate environment?

Having circumnavigated the globe in search of answers to how plant diversity evolves and why ecological processes foster its persistence, the remainder of the book examines the practical implications of these discoveries. In "Liebig's Revenge" we visit the world's longest-running ecological experiment in its 150th anniversary year and discover how and why atmospheric pollution is undermining plant diversity in Europe and North America. Who Liebig was and how he had his revenge you will have to wait to discover.

In chapter 8, entitled "Florida!," we see that plants transported by humans from one part of the globe to another can become invasive, manifesting all the destructive habits of the most fearsome demon. The state of Florida is particularly badly affected by invasive, alien plants but the problem is becoming a global one. Chapter 9 asks whether genetically modified crops may become the new Darwinian demons of the future, or whether the new technology is merely a smarter way to breed new crops. Finally, in "The End of Eden," I consider the bleak prospects for global plant diversity and what might be done to rescue the situation. That is our itinerary, so let's get going!

2

The Tree of Trees

Throughout the fourteen chapters of *The Origin of Species*, there is but one illustration. It is a tree: an evolutionary tree. No other metaphor so compactly and completely sums up what evolution is all about. Tracing branches downward from branch tips to root emphasizes the common ancestry of all life. Following the tree upward from root to branches emphasizes the evolution of diversity, but not all branches lead to the crown of the tree. Some whole limbs have snapped off: they have no living descendants. Others have branched profusely and tens of thousands of living species blossom in the canopy. Darwin believed that representing the relationship among groups of living things in the form of a tree was no mere metaphor, that it "largely speaks the truth":

> As buds give rise by growth to fresh buds, and these, if vigorous, branch out and overtop on all sides many a feebler branch, so by generation I believe it has been with the great Tree of Life, which fills with its dead and broken branches the crust of the earth, and covers the surface with its ever branching and beautiful ramifications.

If Darwin were writing today, a tree would undoubtedly be his logo. The dead and broken branches that encrust the Earth are species known only as fossils, such as the giant lycopods and seed ferns that grew in the swamps during the carboniferous period and later became coal. Limestones such as chalk are the fossil remains of tiny creatures whose shells accumulated over millions of years in marine sediments. It is estimated that at least 95 percent of all species that have ever lived are now extinct.

The species alive today are the outermost branch tips of a very ancient tree. If we want to understand the history of evolution and where present-day biodiversity comes from, we must reconstruct the tree of life. In this exciting scientific enterprise, botanists have been in the vanguard and zoologists, for once, in the guard's van. In its essentials, the green limb of the tree of life, or what poetic licence might permit us to call the "tree of trees," is now known.

Reconstructing the tree of trees has been a formidable technical achievement. The tree metaphor can help us understand just what is involved. It is late autumn as I write, and looking out of the window of my study into our small garden, I can see that the lawn is covered in leaves. There could easily be four hundred thousand of them, a quantity comparable with the number of flowering plant species that exist today. Reconstructing the tree of trees is a problem very similar to trying to place each leaf on the lawn back onto the right tree and onto its original twig. This sounds difficult enough, but now imagine doing it without knowing the shape of the tree! The evolutionary tree does not have the tangibility of wood—it is a genealogy of species. The term used for such a genealogy is "phylogeny." There is no register of births, marriages, and deaths for this kind of genealogy. The only way to comprehend the structure of a phylogenetic tree is to infer it from the relationships among its individual twigs and work backward toward the root of the tree.

Suppose that we had the task of reassembling the trees in my garden with nothing more to go on than a pile of fallen leaves. How would we do it? The first step would be to decide which leaves belonged to which tree. This is the easy part, as there happen to be only three trees: an oak, an apple, and a cherry. Sorting species into their major groups is an ancient science. The Greek philosopher Theophrastus decided that a good classification was to divide plants into herbs, shrubs, and trees. Such a taxonomy would be quite useless for phylogeny reconstruction because plant stature contains almost no information about how closely related any two species are to each other. In the grass family, for example, there are bamboos as big as forest trees, though no one would describe grasses as trees.

Which characters *are* informative about phylogenetic relationships? Sorting leaves by shape will work for separating an oak, an apple, and a cherry tree, but it will not do for a wider range of species because natural selection is forever altering leaf shape to fit the environment. For example, the leaves of the majority of tropical rainforest trees are oval, with a "drip tip" at the end that helps drain the leaf of water and makes it harder for algae and lichens to colonize the leaf surface. The leaves of a group of Australian mistletoes mimic the various leaf shapes of their different *Eucalyptus* tree hosts. This mimicry probably helps the mistletoes hide from butterflies that use visual cues to find their food plants. Thus, convergent evolution makes leaf shape worthless for phylogeny reconstruction: the character does not tell you who is related to whom. This a general problem with all rapidly evolving characters.

Very slowly evolving characters are equally uninformative. With the exception of a few parasites that have lost the ability, all seed plants produce chlorophyll. This green pigment definitively separates plants and algae from other groups, but "greenness" doesn't serve to distinguish among different groups of plants. For obvious reasons, possessing chlorophyll is a character that evolution has carefully preserved from change. What is needed for phylogeny reconstruction is something in between: a character that evolves just quickly enough to differentiate one major group of plants from another, but slowly enough not to permit a mistletoe to disguise itself as a eucalypt.

Botanists have for centuries sought a satisfactory set of characteristics on which to classify plants. In the pre-Darwinian era, this search was motivated by the desire to comprehend the plan of Creation, to fathom the idea in the mind of God. After Darwin, most plant taxonomy was, at least implicitly, a search for evolutionary relationships. Carl Linnaeus, the eighteenth-century Swedish prince of botanists, laid the groundwork for the task of phylogenetic reconstruction that was to come, though Linnaeus himself was not an evolutionist. First and foremost, he established a system of nomenclature that provided a stable foundation for later classification. In his book *Species Plantarum*, published in 1753, Linnaeus described and named all the plants then known to Western science. Many of the approximately 5,900 plants he described still carry the scientific names he gave them. In effect, what Linnaeus did was label each leaf in the pile for us, an essential first step to rebuilding the tree.

Linnaeus grouped his species into genera, genera into orders, and orders into classes. This hierarchical arrangement naturally prefigured an evolutionary tree, though of course Linnaeus did not perceive it that way. He separated plants into classes depending on the number and relative lengths of the stamens in the flowers, dismissing all other systems of classification. This sexual system of classifying plants outraged, titillated, and fascinated the reading public. Samuel Goodenough, later to become bishop of Carlisle, remarked that "nothing could equal the gross prurience of Linnaeus' mind." The German poet Goethe was worried that botanical textbooks that taught the "dogma of sexuality" would embarrass female or youthful readers. Erasmus Darwin, by contrast, was such an enthusiast that he initiated a translation from the Latin original. Typically, Erasmus also devoted half of his book *The Botanic Garden* to a poem titled *The Loves of the Plants*, which reproduced the complete Linnean sexual system in verse. With a panache for rhyme that would do Dr. Seuss credit, he wrote:

> What Beaux and Beauties crowd the gaudy groves,
> And woo and win their vegetable Loves"

before proceeding to describe the full gamut of Linnean sexual arrangements found in different plants:

> Sweet blooms GENISTA in the myrtle shade,
> And *ten* fond brothers woo the haughty maid.
> *Two* knights before thy fragrant altar bend,
> Adored MELISSA! And two squires attend.

Genista has ten stamens (the fond brothers). *Melissa* has two short stamens (the knights on bended knee) and two long ones (the squires that stand to attention). Like Erasmus's poetry, the Linnean sexual system was too artificial to stand the test of time, but both highlighted that the sexual parts of flowers are promising material for plant classification. The structure of flowers is to this day used in plant taxonomy, though modern botanists use all available features of plant morphology and not just the number and disposition of anthers and stigmas.

Classifying plants on the basis of a wide range of structural characteristics, and more recently on chemical ones, brought us a long way from a pile of unsorted species to some rational groupings. By the 1980s, however, the enterprise had reached a stalemate and there were at least three competing views of which plants belonged together on the same branches of the tree of trees. The systems of the American Arthur Cronquist, the Russian Armen Takhtajan, and the Swede Rohlf Dahlgren each had their devotees, and yet there seemed to be no way to pick which was best. To be sure, the systems were in agreement on many points, but there were enough differences to matter. Maybe it would never have been possible to resolve these differences based on anatomical and chemical characters alone. Fortunately, another source of evidence was soon to be discovered, hidden in the genetic code.

The story now takes us back to Kew. From the north door of the Princess of Wales Conservatory we turn right and walk through the grass garden. The flower beds here are filled with tall, exuberant grasses contained within borders formed by other members of the family that are more biddable to the mower. Beyond the grass garden is a water garden with a display of water lilies, and overlooking this is the main laboratory building. Inside, we have an appointment with Mark Chase, a soft-spoken American who is the head of the Molecular Systematics section. Mark came to Kew in 1992, the year before he published the first phylogenetic tree that showed how the majority of flowering plant families are related to one another: the first tree of trees to have all its major limbs sketched in place.

Mark greets us in the lobby and we follow him to his laboratory. Molecular biology laboratories are nothing much to look at, and this one is no exception. Benches divide the main lab into four bays, in each of which there are small pieces of equipment and racks of sample tubes, with figures in a white coats carefully performing repetitive tasks. In another part of the laboratory is the DNA sequencer, looking like the cabinet of a large tabletop freezer, hooked up to a computer. The personnel are international, from Africa, Finland, Spain, Latin America, the United States, each bringing the cadences and accents of their mother tongue to the scientific discourse in English.

The laboratory is like a movie theater. The room itself gives little clue to the excitement that is generated when a switch is thrown, the lights dim, and the celluloid spools through the gate. So it is in the lab. Once a day the DNA sequencer is run, the readout spools from the computer and then comes the excitement. There is a story in the DNA sequence. What is it? What does it tell us about where this plant or that fits among its relatives into the tree of trees? Does it fit where we thought, or is there a surprise branch that we didn't know about before? Mark says that more than once his laboratory has played host to a sceptical plant taxonomist unable to believe the stories the DNA tells. They receive training in the lab, are amazed at the power of the technique, and leave confirmed DNA junkies, sequencing every plant in sight with the zealotry of the convert.

Even more than a decade later, the most significant triumph of molecular plant systematics remains the first "big tree" that Mark published after creating and bringing to fruition a collaboration involving forty-one other scientists. The 1993 paper will probably always be seen as a turning point and it has already attracted the attention of historians of science. The chief secret of the paper's success was that it used characters that evolve at just the right rate, neither too fast nor too slow, to preserve changes that mark the initiation of most of the families and genera of flowering plants. Molecular systematics, as its name implies, uses molecular rather than anatomical characters to make inferences about relationships among groups. The molecule in question is DNA, the hereditary material itself, the stuff of which genes are made.

A gene is a set of instructions to the cell, written in code. The genetic code has just four letters, called "bases." The four bases are the building blocks of DNA molecules and are denoted by the letters A, C, G, and T. A typical gene consists of a piece of DNA containing a string of up to several thousand bases, like an immensely long sentence. The sequence of the bases in the gene—that is, the order in which the four bases follow one another—spells out the instructions it contains, just as the arrangement of letters in the words you are now reading determines their meaning. Rearrange the letters in the phrase "fungal systematics" and it becomes "fantastic ugly mess."

Before cells divide, the genes are replicated so that each daughter cell has a complete set of instructions. This is when mistakes in gene sequences sometimes occur because a base is left out or substituted by another during copying. If not corrected, the change in gene sequence is transmitted to all future offspring in the form of a mutation. A mutation that impairs the function of a gene is normally removed by natural selection, but many genes contain sequences that have little or no effect on function. Indeed, the majority of DNA sequences in our own genome do not provide instructions for anything, so far as we know. These sequences are still replicated and transmitted down the generations. The persistent mutations they contain form an indelible mark of ancestry. Over time, with gene sequences being copied millions of times over, mutations accumulate. As in the party game Chinese whispers, inaccurate repetition can so corrupt the message "Send reinforcements, we are going to advance" that it becomes "Send three and fourpence, we are going to a dance."

Molecular systematics uses the mutations that occur in DNA sequences as characters from which to infer the relationship of species to one another. The logic is simple and based on the idea that two species sharing a mutation that is absent from the original sequence must be related to one another through an ancestor that carried that mutation. It is not just in evolution that copied mistakes can be used as markers to identify the source of information. This is also the surest way to detect plagiarism among students. As in the classroom, so in nature, a single shared mistake may occur by chance, and a couple may be a freak accident. But a battery of shared mistakes or a single very unusual one greatly increases the likelihood that copying has taken place.

Let us now return to the tree of trees to see how shared mutations can reveal its whole structure to us. The first thing that has to be done is to identify a useful gene that can be sequenced in every species that we wish to place on the tree. Mark Chase and his collaborators chose a gene called *rbcL* because a large number of laboratories were sequencing this particular gene in a wide variety of different projects and he thought sequences might be available for about two hundred species. There was, however, a current of opinion at the time that said a dataset of this size would be impossible to analyse. Mark and some of his co-authors "were told by several well-known molecular phylogenists that our efforts were doomed simply because we wished to include so many sequences"; but with encouragement from elsewhere Mark and two other systematists, Doug Soltis and Dick Olmstead, "persisted in the notion that an attempt would be worthwhile and circulated an informal invitation to anyone working on seed plant *rbcL* to send sequences. . . . We were amazed to find that we rapidly accumulated over 400 sequences. At this point, we still had no clear intention to publish our findings."

It seems extraordinary now that such important research findings and the paper that published them started out like this, but we have to remember that no one had analyzed so big a dataset before or attempted so ambitious a reconstruction as the tree of trees. The dataset was so out of the ordinary that the computer program normally used for tree reconstruction had to be specially modified and new routines devised to handle the sheer volume of the data. The final dataset contained five hundred *rbcL* sequences representing the majority of flowering plant families, and some conifers and other species as a so-called outgroup. The outgroup consists of organisms that are known on good evidence not to belong in the phylogeny being reconstructed, but that share a common ancestor with it. The outgroup provides the reference sequence at the root of the tree. It is used to make a best estimate of what the *rbcL* sequence was before the flowering plants diversified.

We can better appreciate why there was scepticism toward the whole enterprise from some quarters if we consider how many possible phylogenetic trees there might be for a group of five hundred species. This phylogeny must have five hundred branch tips. Starting with the first tip, this could be any one of the species, so there are five hundred possibilities straight away. The second tip can be any of the other species, so the possible combinations of species at just the first two tips are 500 × 499, or 249,500. Multiply this by the 498 possible species that might sit on the third branch tip and we get 124,251,000. You get the picture. It looks as though the number of possible trees might as well be infinite. Could the right one ever be found in this universe of possibilities? Indeed, could you even be sure that there wouldn't be thousands of equally plausible patterns for the tree of trees? If there were, this would send Cronquist, Takhtajan, and Dahlgren laughing all the way back to the herbarium for a party.

There was something else to worry about too if the big tree was going to make any sense. The five hundred species in the *rbcL* analysis were to be used as placeholders to represent the very much larger number of species in the angiosperms as a whole. About 350 plant families were represented by one or two species each, but what if families were not natural groups? A few cuckoos in the nest would really confuse the family tree! It was important that families members really did belong together so that placeholders could stand in for each family. In the jargon, each family should be "monophyletic," meaning that all its members must have a unique common ancestor not shared with any other family. At the time of the *rbcL* analysis it was already suspected that the saxifrage family and some others were not monophyletic, so extra representatives of these families were included as a check. With some notable exceptions that need reclassification, the families of plants described by traditional taxonomy—such as the grasses, the daisy family, the rose family, and so on—have been confirmed to be monophyletic.

Zoological nomenclature, particularly in the invertebrates, is in nothing like the fortunate state of botany, and I was once vehemently attacked in public by an entomologist who failed to understand the difference. It happened at Leiden University in Holland, where I had just given a talk about some of my own research in which we had used the 1993 big tree. He stood up at the end and denounced me for "lying with phylogenies." There were ten minutes of mutual incomprehension before I realized what he meant; I was as unaware of the difficulties in insect phylogeny as he was of the successes in plants. I simply failed to understand where he was coming from. His point was that the big tree didn't represent a phylogeny of plant families, only the relationships among the five hundred placeholders that had actually gone into its construction. He was wrong because most recognized plant families *are* monophyletic and therefore each *can* be represented by only one or a few members.

Science and scientists thrive on argument, but one thing everyone *could* agree on was what the *rbcL* analysis should aim for: it was the *shortest* tree. The shortest tree is the one that invokes the least number of separate mutational events that is consistent with the distribution of mutations among the species. This target derives from a general philosophic principle followed throughout science called Occam's razor, which says, in effect, that one should always prefer the simplest hypothesis that is consistent with the facts. The watchword is "parsimony." Easy to state, but not so easy to achieve. Indeed, Chase and his forty-one co-authors finally published their tree of trees knowing that it was *not* the shortest possible tree, but believing that shorter ones would not represent much of an improvement. A leading plant evolutionist at Harvard withdrew his name from the paper in the belief that shorter, better trees would be found. For several years afterward it was possible to check his lab website on the internet to see how many shorter trees his computers had so far come up with. The tally eventually ran into thousands, but the main limbs of the tree identified in the 1993 paper proved robust and more recent research using other genes has strengthened confidence that they are right.

Traditionally, the angiosperms were divided into two groups, the broad-leaved dicotyledons (or dicots) such as oaks and the monocotyledons (or monocots), which have narrow leaves with parallel veins such as in the grasses. The 1993 tree of trees *did* turn out to have two major limbs, but they did not neatly separate the traditional monocots and dicots. The tree showed that the traditional dicot characteristics had evolved early in the history of the angiosperms, before the origin of the monocots. Where the tree of trees divides near its base, one limb bears the majority of living broadleaved species that have now been given the name "eudicots," but the other carries a mixture of some broad-leaved plants and all monocots.

The anatomical feature that separates the two major limbs is not in their leaves but in the microscopic structure of their pollen. Eudicot pollen grains have three apertures, whereas those of the monocots and broadleaves on the other limb all have a single, grooved aperture. This latter type of pollen seems to be the ancestral one, from which the eudicots evolved their three-apertured type. Whether the evolution of a novel kind of pollen itself gave the eudicots a major advantage, or whether it is merely a fortuitous indicator of something else of importance, is a mystery. What we do know is that this change marks the origin of a lineage that gave rise to three-quarters of all the kinds of flowering plants alive today. Pick a species at random, and three times out of four it will be a eudicot. Roses, oaks, heathers, daisies, violets, buttercups, cabbages, and nearly two hundred thousand other plants belong to this group. It is a demon lineage.

In 1998 Doug Soltis, one of the original gang of three behind the big *rbcL* tree, Mark Chase, and Vincent Savolainen put together a dataset with sequences for two more genes, making a computer file three times bigger than had been used for the *rbcL* tree. One day, Doug and Pamela Soltis, who work closely together as a husband and wife team, set their computer running and then left the lab for the night. Such a big dataset clearly wasn't going to yield an answer while you waited. (The Harvard computers would probably still be looking for shorter trees in the one-gene *rbcL* dataset if someone hadn't pulled the plug.) Next morning, Doug and Pam made their way to the lab, expecting their data still to be running. Who knew *when* the program would finish the search? There was a surprise: the program had stopped. There was no obvious error, as the program had apparently converged on an answer. It seemed to have found the shortest tree, but surely this had to be a mistake? How could three times as much information be processed in a fraction of the time it took to get even an approximate answer out of *rbcL?* Doug double-checked the data file but could find nothing wrong, and so he ran it again. The same result, just as quickly. This was a startling discovery. The more data you put in, the quicker and more sure the result! Could this be true? It turned out that it was. The reason why is easy to understand once you know the answer, but no one expected it. More evidence of relationship means fewer blind allies to search, and so a faster route to an answer.

The main limbs of the 1993 tree have been successively confirmed by a two-gene tree, a three-gene tree, then a five-gene tree, and at the time of this writing a ten-gene tree is in the works. With each addition to the evidence more of the smaller branches and knottier relationships that the 1993 tree left unresolved have been cleared up. The sceptics are now converts. Curiously enough, the sceptical reception that the idea of a tree of trees received at the end of the twentieth century echoes some of the criticisms that were made of the concept of the gene at the century's beginning. Leading geneticists, notably William Bateson at the University of Cambridge, believed that the gene was a useful concept, but not an actual "thing." The discovery of the molecular structure of DNA by Jim Watson and Francis Crick in Cambridge in 1953 finally nailed that one. The structure of the tree of trees is also turning out to be a lot more attainable than the sceptics thought.

Every six years the botanists of the world gather for an International Congress and in 1999 the Sixteenth International Botanical Congress met in St. Louis, Missouri. Plant systematists hold a nomenclature session before each congress to agree on changes to plant names in the light of new research. These only upset ecologists like me who would rather not be told that bluebells, which have already been moved from the genus *Endymion* to *Scilla*, now belong to *Hyacinthoides*. So, by the time I arrived in St. Louis the advance guard of botany had already pitched their tents and were in a state of high excitement. Word was out that the three-gene tree had just thrown up the answer to an age-old mystery: which is the lowest living branch in the tree of trees? The answer was something called *Amborella trichopoda*, a rare and lonely plant from New Caledonia, the only member of its family still in existence. It was so rare no one could exhibit a picture of it; the single photograph of the plant that seemed to be at the congress had been loaned to the press office for use by journalists. Credibility on the streets of St. Louis that week depended on at least being able to remember the name of this celebrated and hitherto unheard of (by me) plant. I solved this problem with a mnemonic inspired by *Mary Poppins:* "Any Amborellas, any Amborellas to mend today?" (Don't try it! You'll never get the damned tune out of your head.)

Down there at the base of the tree of trees, one rung up from *Amborella* and older than the eudicots, is a much more familiar group of plants, the water lilies, the family to which the giant *Victoria amazonica* belongs. Not long after the molecular systematists showed that water lilies branch near the bottom of the angisoperm phylogeny, fossil evidence was discovered that supported the ancient status of this group. A tiny, three-millimeter-long water lily flower was discovered in deposits from the early cretaceous period in Portugal. What a small beginning had the three-hundred-millimeter flowers of *V. amazonica!* The tiny fossil showed the water lilies to be at least 115–25 million years old, placing them among the oldest fossil angiosperms. I wonder, did a wry smile pass across Mark Chase's face when the basal origins of the water lilies were discovered and he saw them blooming in the water garden, right beneath his laboratory window at Kew?

Fossils and phylogenies are complementary sources of historical evidence. Neither provides us with the whole picture of evolutionary history on its own, but when they corroborate one another the result is compelling. Fossils tell us what extinct plants looked like and various dating techniques can be used to determine the age of fossil finds. But, there are gaps in the fossil record because organisms must die in rather unusual circumstances to become fossilized. The normal fate of the dead is to become food for a nether world of organisms that feast on them. Only in conditions where microbes and other consumers cannot live—because it is too arid, too cold, or so waterlogged that there is insufficient oxygen—do remains stand a chance of preservation. Thus, many species have left no fossil record of their existence.

By contrast, molecular phylogenies are built from gene sequences of living organisms. One day, all these groups will be placed in their correct positions on the tree of life. Though molecular phylogenies are silent about extinct species, they tell us a good deal about the history of living groups. This information comes from the structure of phylogenetic trees and, in particular, from the lengths of their branches. The length of a branch is measured by the number of mutations that separate the sequences at either end. At the branch tip sits a living species and at its base is the common ancestor of that branch and the sister branch. Mutations are relatively rare events and their accumulation requires the passage of time. Changes in DNA sequence can therefore, at least in principle, be used as a molecular clock. A short branch with few sequence changes implies a more recent origin than a long branch with many.

A difficulty is that the molecular clock in *rbcL* and other plant genes does not run at a uniform rate. It runs more slowly in the palms than other plants, for example. A dozen base changes between two palm sequences represents a bigger gap in time than the same number of changes in the rose family. Nonetheless, there is a striking pattern that occurs time and again in the tree of trees: the majority of plant families sit in splendid isolation on the end of long branches. The tree of trees looks like the *Aloe barberae*, that aloe-on-a-stick in the Princess of Wales Conservatory at Kew (chapter 1). It has numerous long naked limbs that sprout only at their ends. What does such a pattern mean? Even if the molecular clock runs at a different speed in different lineages, the difference is only one of degree and the conclusion is the same for all of them. In the tree of trees a long naked limb indicates that a family has an ancient origin, whereas the sprout of short branches at its top means that its genera and species are relatively recent. The modern biodiversity of living species and genera has a recent origin, but the family roots run deep in geological time. This pattern has profound implications as well as deep roots.

First, the pattern explains why the majority of the plant families, many of them recognized by botanists even before the time of Linnaeus, fitted so well into the molecular phylogeny. Families are real evolutionary groups that are distinct enough from one another to be easily recognized, and they can be distinguished in all kinds of ways. This is why the tree of trees built using placeholders is not a phylogeny of lies.

Second, there is an evolutionary mystery implicit in the structure of the tree of trees. Why do so many families contain just a single, ancient lineage with no branching below the top? There must once have been branches on these limbs, but what happened to them? There is a clue to the answer in a pattern seen in the grasses and their sister family, the Ecdeiocoleaceae. There are about seven thousand grass species, but Ecdeiocoleaceae has five times as many vowels to its name as species. This pattern repeats itself time and again in other sister families: the family on one branch has ramified and is very species-rich, but its sister family appears to be on the way out. Surely there are shades of the Darwinian demon here? Do successful lineages expand at the expense of similar but less well-adapted relatives? A case of "This planet ain't big enough for the two of us"? Are side branches shaded out by the success of the main family stem?

Darwin knew nothing of the structure of the tree of trees, but his words from *The Origin* that I quoted at the beginning of this chapter now resonate with additional strength. Recall that he wrote that a branch on the tree of life would "if vigorous, branch out and overtop on all sides many a feebler branch." Darwin's insight continues to astound and illuminate, a century and a quarter after his death. The message is clear: Darwin may be dead, but his demons are alive. But, how then, in a world threatened by demons, does diversity evolve?

3

Succulent Isles

A hundred kilometers from the coast of Morocco, where the Atlantic shoulder of North Africa rounds toward Spain, lies the volcanic island of Fuerteventura. One hundred kilometers further west into the Atlantic slumbers Gran Canaria and a similar distance westward again is Tenerife. From there, three more westward volcanic islands complete the reach of the Canary Islands archipelago into the Atlantic. The Mediterranean, almost African, climate of the Canaries attracts tourists from throughout northern Europe. There is, however, another reason to visit: the islands are a paradise for plant lovers in search of novelty and a "must" for anyone who wants to see evolution in action. Perhaps most spectacular of the plants are the dragon trees, a species of *Dracaena* that many people grow in their homes, but in Tenerife there is a wild specimen so large that it has a door fixed into its trunk, giving another meaning to the term "houseplant."

Darwin called here on January 6, 1832, ten days after leaving England on the outward leg of his journey aboard *HMS Beagle*, but he never set foot ashore because the ship was quarantined by the authorities in Tenerife for fear that it might be carrying cholera. Had he landed, one wonders, would the flora of the Canaries have inspired Darwin's theory in the way that the fauna of the Galapagos archipelago eventually did on the other side of the globe? He certainly could not have failed to notice the great variety of succulent plants that occur only here—like the weird and wonderful houseleeks (*Aeonium* species) that evolution in the Canaries has fashioned into a wide range of growth forms, from rosettes flattened to the rocks like upturned rubber saucers to mops of leaves on branched, meter-high sticks. Such evolutionary novelties have earned the Canaries the *soubriquet* "the Galapagos of botany." Archipelagos like the Canaries and the Galapagos are species factories where newly forged varieties spill abundantly from the cornucopia of evolution. The reason that the evolution of new species is more obvious on remote islands than elsewhere is because they have usually acquired their flora and fauna quite recently. On continents, the signs of recent speciation are harder to find because these are old landscapes that evolution has diligently stocked for hundreds of millions of years. Countless demons have wrestled with each other across these landscapes and have had time to hone their arts of ecological dominance, leaving very little room for new species to colonize. The Canaries and the Galapagos are different because they are geologically new and biologically isolated.

All these islands were born at sea, far enough from any continent to avoid being deluged by species as soon as the lava of their formation had been quenched by the ocean. Because plants and animals had to cross large distances to reach the islands, they filtered onto them one or two at a time, only to find mostly empty habitat. What happened to them next is what interests us. What can the succulents and other endemic plants that evolved from these colonists in the Canaries tell us about how and why diversity evolves?

About 40 percent of the native plant species found in the Canaries are endemic, with the remainder of the flora composed of species also found in Africa or the Mediterranean. The dragon tree, for example, is African. Because the majority of the Canarian flora is shared with the adjacent continents, it was believed until quite recently that many of the endemic species in the Canaries were probably just the surviving relics of populations that once existed in Africa or the Mediterranean, but that became extinct there when the climate of those regions became drier. The laurel forests in the mountains on the higher islands are home to trees that resemble fossils from forests that once grew around the Mediterranean. If all the Canaries' endemics were just relics, however, the archipelago would be more accurately described as a museum of near-extinct species than as a species factory. This would not diminish the botanical interest of the endemics, but it would mean that they would have more to tell us about the demise of diversity than about its efflorescence.

Recent research has found that, in almost every case so far examined, the plant groups containing the largest numbers of Canary Island endemics evolved and diversified in the archipelago itself. These species are therefore not mere relics of a lost continental flora. The story is another triumph of phylogeny reconstruction using the kinds of methods described in chapter 2. The Canary Island subspecies of the olive tree (*Olea europaea*) provides a good example. An analysis of DNA from populations of the olive from Africa, the Mediterranean, and the Canary Islands suggests that the closest living relative of the Canary Island subspecies is to be found in Morocco. Birds probably carried olive fruit from there to Fuerteventura, the nearest of the Canary Islands and also, at about 21 million years, the oldest. The DNA evidence suggests that the olive then island-hopped its way westward along the archipelago to Gran Canaria (14 million years old), from there to Tenerife (about 7 million years old), and then La Gomera (12.5 million years old), finally reaching La Palma, which is a mere 1.5 million years old.

Precisely when the olive first colonized Fuerteventura is not known, but there were certainly pauses between each successive step along the archipelago because each island population had long enough to evolve its own peculiar genetic hallmark before this was then transmitted by dispersal down the line to the next island. The phylogenetic relationships among island populations and the pathways of dispersal were worked out from the trail left by these genetic hallmarks. The use of molecular phylogenies to reconstruct the migration pathways taken by populations in the past is known as "phylogeography." It works on the assumption that species sitting on the lowest branches of a phylogenetic tree are not only genetically close to the ancestor of the group, but also geographically close to the location of that ancestor. Thus, the olives of Fuerteventura sit on a branch that is sister to Moroccan olives and must therefore share an ancestor with them from that part of Africa. Each of the olive populations from the other Canary Islands is descended, directly or indirectly, from the population on Fuerteventura and occupies a progressively higher branch in the phylogeny, with the youngest population from La Palma at the top.

What is truly extraordinary about this pattern is that olives from all the different Canary Islands share a common ancestor: they all belong to one and the same lineage that can be traced back to the colonization of Fuerteventura by an olive from Africa. This is extraordinary because it means that the olive successfully colonized the Canary Island archipelago only once in its 21-million-year history, though quite when we do not know. The more one thinks about this singular event of colonization, the odder it seems. To see just why it is unexpected, let's consider two extreme, alternative scenarios for the arrival of the olive in the Canaries and see if either makes sense.

The first scenario is one of early arrival: assume that the olive reached each island soon after its emergence from the Atlantic, and thus the tree has been on each island for about as long as it has been habitable. This would mean that the olive made the one-hundred-kilometer leap to Fuerteventura about 21 million years ago and then leapt another hundred kilometers to Gran Canaria 7 million years later and the same distance again to Tenerife 7 million years after that. This gives a dispersal rate of three successful crossings of a one-hundred-kilometer distance in 14 million years (one to Fuerteventura, one from there to Gran Canaria, and then to Tenerife), or about one and a half colonizations per 7 million years. On this basis, allowing for the fact that a half colonization cannot occur, we might expect Fuerteventura to have been successfully colonized a total of four or five times in its 21 million year history, Gran Canaria three times in 14 million years, and Tenerife once or twice. The total number of expected colonizations for the three islands is therefore between eight and ten. However, the phylogeographic evidence tells us that the actual rate of successful dispersal was only three, which is only three-eighths to three-tenths of the expected frequency. These back-of-the-envelope calculations show that the early dispersal scenario does not seem to fit the facts. What, then, about an alternative scenario—recent dispersal? Does it do any better?

If we assume that the entire colonization of the Canaries happened after the youngest island where it is found, La Palma, emerged from the sea, then we have the strange situation in which the olive was unable to successfully colonize Fuerteventura during the first 19.5 million years of its history, but that in the remaining 1.5 million years it then jumped one-hundred-kilometer distances three times. This raises two questions. First, why was there such sudden success in dispersal after so long a period of failure? Second, if the rate of dispersal really has been so unusually high in recent times, why has the olive successfully colonized the Canary archipelago only once in the last 1.5 million years?

In fact, neither of the two dispersal scenarios really seems to be able to explain why the olive colonized the Canaries only once. Intermediate scenarios, somewhere between the extremes we have considered, are no better. When facts don't seem to fit, one should always consider the likelihood that they are mistaken. Could it be that the phylogeny of the Canary olive is wrong and that there have been multiple colonizations? Such a mistake could have been made through insufficient sampling of island populations, for example. A good check is to see whether results can be repeated. In this case, the result of the check is quite clear: molecular phylogenies for other Canarian endemics show that many of them also colonized the archipelago only once. The thirty-nine *Aeonium* species, for example, all have a common ancestor in the archipelago. A group of thirty-four Canarian endemics related to the sow thistles evolved from a single ancestral colonizer belonging to the sunflower family—a pattern repeated in at least four other Canary Island endemic groups in the same family. These and other examples suggest that singular colonization by the ancestors of the most species-rich endemic groups is a rule that is rarely broken in the Canaries. Why? The solution to this singular mystery leads to an explanation of that bigger puzzle: how diversity evolves when demons threaten.

Let's think about the colonization process a bit more deeply. Successful colonization of an island depends not only on getting there, but also on finding conditions that permit establishment and spread on arrival. In other words, not one but two barriers must be crossed: the first is the sea, but the second is on land. Our back-of-the-envelope calculations showed that the difficulty of overcoming the obstacle of distance cannot really explain why colonization occurred only once. The idea that dispersal over a distance is not the main barrier to successful colonization is supported by the fact that, in addition to the ancestors of all the endemics, some six hundred nonendemic plant species have also successfully colonized the Canaries. So, dispersal itself hardly seems a rare event at all!

If the explanation for singular colonization is not that the probability of olives, houseleeks, or sow thistles reaching the Canaries is low, it must depend on what happens to would-be colonists *after* they arrive. What if colonization over a distance is like a race in which the prize of island possession goes to the first past the post and the winner takes all? Imagine the scene. One day, millions of years ago, a small flock of Moroccan pigeons that has been feeding on olives and is carrying some olive stones is blown off course by a gale and lands, exhausted, on Fuerteventura. There is little or no vegetation on the island and the pigeons soon die in such inhospitable surroundings. The storm abates, but the rain that has fallen is still cascading in sheets down the naked, rocky mountainside into the sea and the torrent seizes the pigeons' carcasses and carries them away—but not all of them, as some lodge fast in rock crevices. There, the seed carriers rot and deposit their loads, fertilizing the first olives' winning posts with their remains.

Fast-forward a few decades: most of the olive trees have not survived, but a handful of the new immigrants, those with particularly hardy constitutions, have made it and are now flowering. These hardy survivors cross-fertilize with each other, and at season's end they bear fruit. Back in Morocco, the fruit would have been carried off by hordes of birds and most of the stony seeds would have been gnawed and eaten by rodents. But here there are only a few birds to take the fruit and no mammals at all to gnaw the seeds. The result is that the ground around the trees is saturated with rotting olive fruit containing viable seeds. A year later, the first generation of native-born Canarian olives germinates and every nook and cranny thereabout becomes home to a seedling olive tree. Thirty years more, and Fuerteventura has an olive forest big enough to attract a passing flock of pigeons to feed on its fruit. Now, with no shortage of seed dispersal agents and a complete absence of seed predators, there is an olive population explosion and soon everywhere on Fuerteventura that an olive tree can grow has been colonized by a descendant of the first olives to arrive.

So great is the fecundity of most plants and the power of geometric increase that a scenario such as we have imagined could be completed within just a few generations of the first seeds arriving. And, once complete, with every spot that an olive can grow already occupied, the next dead pigeon flock to wash up on the island would deposit a seed load with no future. If true, a scenario like this could explain the once-only rule of island colonization. The explanation has very little to do with the frequency of dispersal and everything to do with the ability of the first colonizers to preempt all habitable space before the next colonizers arrive. In effect, the first colonizers bolt the door behind them. But, you may be asking yourself, isn't this a mere just-so story that proves nothing? And worse, if it is true that demon colonizers inhibit later arrivals, how do we get so many species on islands? Both questions have an answer.

The pigeon story is, of course, imaginary, but its two most important biological details are established fact. First, we know that the olive did make it to Fuerteventura. Were you to visit the island today, you would find that wild olive trees are very rare and that Fuerteventura is more desert than forest. There is good evidence, however, that human colonization of the island is responsible for this, particularly the introduction of goats, which have destroyed the native vegetation of many an oceanic island around the world. Ancient olives trees and other forest plants can still be found in inaccessible places on the cliffs of Fuerteventura—relics of the vegetation that probably covered the island when it was in its pristine state.

Second, we know that all populations have the inherent, demon ability to multiply rapidly when unchecked by competition or predation. One narrative detail that might be questioned is the absence of mammalian seed predators on Fuerteventura. There are almost no native land mammals in the Canaries today, but the remains of recently extinct giant rats have been found on Tenerife and Gran Canaria. If such animals were ever present on Fuerteventura they could have eaten the first olive seeds, slowing the spread of the tree. However, it is just as likely that giant rats could have aided the spread of olives across the island since most rodents, as squirrels do, carry away and bury a proportion of the seed that they collect and some of these are not retrieved before they germinate.

Every species must have its day as a demon, spreading uncontrolled for a while, or it would never become established. We also know, thanks to phylogeography, that many species-rich groups of island endemics improbably descend from just a single colonization event. It does not seem at all far-fetched to put these two facts together and conclude that the first is the explanation for the second and that demonic colonizers inhibit the establishment of later immigrants.

Now we come to the really knotty question. By explaining how demons limit the colonization of islands, haven't we solved one puzzle only to produce a much more difficult one? The problem is the familiar paradox, and the theme of this book introduced in chapter 1, of how to reconcile the evolution of diversity with the fact that natural selection favors individuals with demon traits and dominating proclivities. If every new species to colonize the Canaries rapidly occupied all the space suited to its needs, how did the new, endemic species that evolved so many times from these colonizing ancestors ever get started? The ecology and phylogeography of Canarian endemics holds the answer.

The Canary Island archipelago belongs to a larger group of islands known as Macaronesia. South of the Canaries is the Cape Verde archipelago; north lie the Salvages, Madeira, and the Azores. These islands are today separated from the Canaries by hundreds of kilometers of open sea, but there are at least fourteen submarine mountains that are near enough to the surface to have provided island steppingstones between the Canaries and the rest of Macaronesia when sea level was lower than it is now. Many species endemic to the Canaries have relatives, and even descendants, in other parts of Macaronesia, suggesting that the now submerged islands were once used in a game of Macaronesian hopscotch. A medal for this sport should be awarded to daisies in the genus *Argyranthemum* whose closest living continental relatives are oxeye daisies. The White Marguerite, familiar to gardeners, is an *Argyranthemum* species.

The entire *Argyranthemum* genus is endemic to Macaronesia where its history of island-hopping can be read in its molecular phylogeny. This shows that the twenty-three *Argyranthemum* species descend from a single colonization of Macaronesia, with one branch of the tree diverging to colonize Madeira and its nearby islands and another, much bigger, branch supporting all the species that evolved from a single solitary colonization of the Canary Islands. In the jargon used by builders of evolutionary trees, each of these major lineages is a "clade," from the Greek word *klados*, meaning a branch. The defining characteristic of a clade is that all it members, however few or many they may be, share a common ancestor. The biggest clade of all, therefore, is the tree of life itself, though usually the term is reserved for its parts.

The branching structure of the Canary Island clade of *Argyranthemum* gives us a fascinating glimpse into how evolution turned one founding species into many. According to the principles of phylogeography, *Argyranthemum* probably first colonized either Fuerteventura or Tenerife, since natives of these two islands branch lowest from the phylogeny. It is not clear from the phylogeny which of the two islands was colonized first, though it was probably Fuerteventura because it is so much the older and is also much nearer the continent than Tenerife. Whichever island it was, *Argyranthemum* leapt from one to the other and this seems to have been accompanied by a radical change of habitat that had far-reaching evolutionary consequences.

Only one species of *Argyranthemum* lives on Fuerteventura. This species, *A. winteri*, is found in humid lowland scrub at an altitude of four hundred to five hundred meters, where trade winds from the northeast bring moisture and rain. By contrast, many species of *Argyranthemum* are found on Tenerife, with different ones in different habitats. The species branching lowest in the phylogeny, and therefore the one that is probably most like the progenitor that first colonized Tenerife, is the eponymous *A. teneriffae.* This species lives in high altitude desert habitat, above two thousand meters, where the trade winds have no moisturizing influence on the climate. Thus, the first two species to appear in the evolutionary history of *Argyranthemum* have quite different ecological requirements. The one on Fuerteventura grows in mesic (nondesert) conditions, while the one in Tenerife is a denizen of the desert.

The first plant from a mesic environment that managed to invade the margins of a desert habitat (assuming it happened that way) must have been unusually drought tolerant. Natural selection would have built on this small beginning and honed the drought tolerance of the invader's descendants with each new generation, enabling them to push farther and farther into the desert and away from resemblance to their ancestors. The spawning of new species by such a process is called "adaptive radiation." The essential point about this mode of speciation is that adaptation to new ecological conditions unlocks access to new resources that make new ways of life possible.

It just so happens that the influence of the northeasterly trade winds and the moisture they bring is so critical that it divides all the habitats of the Canary Islands into two distinct camps: those exposed to moisture-laden winds and those not. Soon after the common ancestor of the *Argyranthemum* genus arrived in the Canaries, natural selection discovered these two fundamentally different kinds of habitat and produced a new species in each of them. That, however, was just the beginning. The phylogeny of the species in the genus divides into two clades, one sprouting from *A. winteri* in Fuerteventura and the other from *A. teneriffae.* Though the two clades originate from different islands, the species that evolved in them are not confined to these islands, and members of both clades are found throughout the

Canary archipelago. This is further evidence that dispersal is commonplace, but it also shows something even more important: the crucial difference between the two clades is not geographical, but ecological.

The clade rooted in Fuerteventura contains species that are mainly found in mesic habitats such as humid lowland scrub, heath and laurel forest while nearly all the species belonging to the clade derived from Tenerife live in desert conditions. By adapting to the two major types of environment in the Canaries soon after its arrival, *Argyranthemum* unlocked the door to every habitat and every island in the archipelago. From there, new habitats and new islands were colonized and in them and on them further new species evolved. Twenty three species are recognized in the genus, but there are also many more incipient species that taxonomists describe as subspecies. Given time, the subspecies are likely to diverge from their ancestors sufficiently to warrant recognition as species in their own right.

The evolution of *Argyranthemum* in the Canaries illustrates two of the chief mechanisms by which new species arise: by colonizing new habitats and by the founding of new populations in remote places. The crucial thing about both of these mechanisms is that they create a barrier that isolates new populations from existing ones so that little or no mating can occur between them. This reproductive isolation between an incipient species and its ancestral population is vital because it lets the incipient species evolve in new directions. Plants that manage to invade a habitat that is different from their own tend to evolve, and eventually to become new species, because life in different conditions exerts new demands that favor novel individuals, though reproductive isolation is required for this to happen.

That new species seem to appear in new environments tells us something fundamental about evolution. In fact, it goes to the very heart of the origin of species. Darwin described in his autobiography how he at first overlooked "the tendency in organic beings descended from the same stock to diverge in character as they become modified." This is the demon question over again: why does evolution produce diversity? If every new species demonically replaced an existing one, the total number of species would not increase through evolutionary time. But it does. Darwin described how the answer to this question came to him:

> I can remember the very spot in the road, whilst in my carriage, when to my joy the solution occurred to me; and this was long after I had come to Down. The solution, as I believe, is that the modified offspring of all dominant and increasing forms tend to become adapted to many and highly diversified places in the economy of nature.

What Darwin described as an organism's "place in the economy of nature" we now call its ecological "niche." To a plant from a mesic habitat, a desert provides a new niche. Adaptation to new niches was Darwin's solution to the problem of how evolution produces diversity. The fact that he deduced the idea without the modern phylogenetic evidence that proves him right is yet another instance of the power of his theory of evolution.

However, there is another way in which new species can evolve. They can also arise by geographical isolation alone, even when there is no ecological change. When plants of a desert species colonize desert on a new island, for example, mere physical and reproductive isolation from their source population can lead to divergence, and ultimately to island speciation. One reason why island speciation happens is because the founders reaching an isolated island will usually be very few in number. Small numbers of individuals taken from a variable source population are bound to be a biased and unrepresentative sample. These eccentric founders pass their peculiarities on to future generations and characters that were unusual or extreme in the source population become the norm in the new one. This is known as the "founder effect." Another way that isolated populations can diverge from their source sufficiently to become separate species is if their numbers remain low for several generations after they have arrived, because this tends to permit eccentrics to multiply by pure chance.

So, start with the founder effect, add some random eccentricity, keep this simmering in small, reproductively isolated populations over a few generations, and you have the recipe for a new species to evolve by geographical isolation. No adaptive radiation into new habitats is needed to kick off this process, but if there do happen to be any ecological differences between old and new populations, natural selection will tend to produce adaptive changes in the new population that will accelerate the rate at which it diverges from its ancestors. If old and new populations occupy different islands, you have island speciation.

We now have an answer to the question of how new species evolve and how diversity arises, even when every successful new species is a potential demon that is capable of suppressing any further speciation. The answer is that new species evolve from the refugees that escape from the demons' dominion. There are two ways to escape: into kinds of habitat that demon ancestors cannot tolerate, or on to islands that they have not colonized. Escapees that colonize new kinds of habitat evolve into new species by adaptive radiation. In addition, escapees that colonize new islands can also form new species through geographical isolation. The two modes of speciation reinforce one another, as in *Argyranthemum* that evolved new species in mesic and desert environments on separate islands, and was then able to invade other islands via habitat types to which the colonists were already at least partially adapted.

As an explanation for the paradoxical evolution of diversity nothing could be simpler, but is it true? Can escape from Darwinian demons really explain the diversity of Canary Island endemics? If it is right that plants must colonize new habitats to escape, then one would expect islands with a greater variety of habitats to have more endemic species on them than islands with fewer habitat types. To check whether this is so, we need an inventory of Canary Island habitats and a local botanist to show us the endemic plants in them.

Our guide is Arnoldo Santos-Guerra, Director of Research at the *Jardín de Aclimatación* in La Orotava, Tenerife, which is the second-oldest botanical garden in all of Spain. Tenerife was once the first port of call for Spanish galleons returning with looted riches, including plants, from the Americas. Arnoldo has to divide his time between the needs of the garden's exotic plants and the exploration and documentation of the region's wild flora, but he spends every available moment in the field. His extraordinary field knowledge was essential to the construction and interpretation of the molecular phylogeny of *Argyranthemum* for which the laboratory analysis was done by another native Canarian scientist, Javier Franciso-Ortega, then working at the University of Texas and currently on the faculty at Florida International University in Miami.

The different habitats found in the Canary Islands are stacked vertically up their volcanic slopes, like the floors in a department store each specializing in different goods. Because Tenerife is the tallest island, it affords the full range of seven habitat types. Arnoldo's four-wheel drive is our express elevator to the top, where the snow-capped peak of Mt. Teide towers over a moonscape of broken volcanic rocks in an unexpected variety of colors so intense that they appear to be fragments of solid pigment: steel black, burnt umber, red ochre, yellow ochre, verdegris, and pumice white. As we follow the road that snakes across this awe-inspiring desert, Arnoldo gives a cry of recognition and pulls the van over. We leap out and gather round to see a specimen of *A. teneriffae*, the oldest species in the dry-habitat clade. From the ancestors of this plant sprang most of the *Argyranthemum* species that inhabit dry habitats on Tenerife and on other Canary Islands.

As we descend the mountain to two thousand meters we enter a belt of Canary pine forest. The immensely long, needlelike leaves of the Canary pine (*Pinus canariensis*) precipitate moisture from the air, causing water to drip from their ends and into the soil where pine roots can retrieve it. The self-watering habit of these trees is a wonderful adaptation to an arid environment that, without the pines, would be unable to support other endemic plant species that live on the forest floor. Two species of *Argyranthemum* belonging to the desert clade live here.

Continuing our ear-popping descent, we round a hairpin bend and suddenly the pine trees by the side of the road open, as if drawn apart like the blinds at a window, and reveal a spectacular view beneath us. We are just above a layer of cloud that the wind is driving against the mountainside, where it billows upward, boils, and then vanishes into the atmosphere. We are witnessing the permanent encounter between land and the cool, humid trade winds from the ocean that create a broken ring of cloud suspended around the north-facing slopes of each island in a belt between four and twelve hundred meters.

The trade winds feed life-giving moisture to three more habitat zones. Highest is a belt of heathland inhabited by an *Argyranthemum* species belonging to the mesic clade; below this is a laurel forest that has been colonized by its next of kin. Laurel forests are almost permanently shrouded in cloud and draw moisture from the atmosphere, which condenses on their waxy leaves and drips from their tips. So important are the water-intercepting powers of trees on these volcanic islands that legend has it that in pre-Spanish times one particular tree on the small island of El Hierro was the source of enough water to sustain all the local inhabitants. The story goes that the native people hid the tree from the Spanish invaders by covering it with dry grass, but that a young girl who had fallen in love with a Spanish soldier revealed the vital secret of their water supply and was condemned to death by her village for the betrayal.

Beneath the laurel forest is a narrow zone of humid lowland scrub where there are no fewer than three *Argyranthemum* species, two belonging to the mesic clade and one to the desert clade. The presence of the latter is perhaps due to the proximity of a belt of arid lowland scrub. This is a dry habitat but it contains one *Argyranthemum* species belonging to the mesic clade as well as another belonging to the desert clade. The mixed phylogenetic origins of *Argyranthemum* in these two habitats at the interface between mesic and dry environments is evidence of adaptive radiation on a very local scale, with colonists from each of the two kinds of scrub able to successfully establish new species just across the border in adjacent habitat. The seventh habitat type on Tenerife is coastal desert, which skirts the entire ocean fringe of the island. Three *Argyranthemum* species are found here, all belonging to the desert clade. I ask Arnoldo how he felt when he first saw the evolutionary tree for the *Argyranthemum* species he knows so well in the field. "For me, it was like a novel," he says. "Chapter 1, Chapter 2, Chapter 3 . . . ." He already knew the ecological roles of each species but, until then, not their evolutionary story.

In all there are at least thirteen *Argyranthemum* species or subspecies on Tenerife. The pattern in the Canary archipelago as a whole is that the taller islands have the most habitat types and consequently the most species. Gran Canaria, for example, reaches nearly two thousand meters and supports six of the seven habitat types in which are found eight species of *Argyranthemum*. Fuerteventura is actually slightly larger than Gran Canaria in land area, but it is low-lying and reaches an altitude of only eight hundred meters. This island has only three habitat types and only a single species of *Argyranthemum*. Thus, as predicted, the number of habitats on an island really does determine how many endemic species have established there. Other factors that one might have thought would be important, such as the total area of the island, its geological age, or its proximity to the mainland are not nearly so important. This applies not only to the genus *Argyranthemum*, but also to the number of endemics as a whole. Sixty-nine Canary Island endemics occur on Fuerteventura, but there are two hundred on Gran Canaria with its greater diversity of habitats. These figures decisively confirm the prediction of the demon-escape hypothesis.

The Canaries are fascinating in themselves, but we visited the archipelago to find an answer to a question of much wider relevance: how does diversity evolve in the presence of demons? The answer turns out to be that it evolves when plants are able to escape by colonizing new territory or invading novel habitats. This escape can trigger profuse speciation that produces clades with lots of short, stubby branches on the end of a long stem. Such a pattern of branching is remarkably like the phylogenies of entire plant families (see chapter 2). Is the resemblance mere coincidence? Or do families also originate from adaptive radiations, each triggered by the evolution of some key innovation or invasion of new territory? Very likely they do, and if so the influence of demons shaped plant evolution in a fundamental way, not only by shearing twigs from the longest stems in the tree of trees as Darwin imagined, but also by providing an impetus that spurred the evolution of plants that were able to escape their clutches—the origin of present-day plant diversity.

In the story so far, the Darwinian demon has been a shadowy, skulking figure whose past influence we have inferred mainly from circumstantial evidence. We have gauged its strength in the power of geometric increase that all living things possess; we have credited the efflorescence of species in new terrain to the absence of suffocating demons; we have detected sudden release from its suppressive presence in the phylogenies of plant families and in the phylogeography of island endemics. But what of contemporary demons? What are their habits and their haunts and, most important of all, what halts them in their tracks? To find out what puts the brakes on demons, we must hunt them where they are active and reign supreme—on Demon Mountain.

4

Demon Mountain

Mount Shimagare, in Japan, is fertile territory for demons and we are here to test their strengths and probe their weaknesses. It's not all demons, however. Meadows on the slopes of these northern Yatsugatake Mountains of Honshu are abuzz with dragonflies and overflow with wild flowers. Some are familiar to us as garden plants, such as gentians, carnations, hostas, and rhododendrons. A more sinister acquaintance lurks here too, for this is also the native habitat of Japanese knotweed (*Reynoutria japonica*), a pernicious invader that reduces seminatural habitats in Britain to simulacra of the barren craters on Mount Fuji where only knotweed reigns.

Japanese ecologists have been studying the forest on this mountain for over half a century and understand it well. There are forest plots in the sub-alpine zone where every tree is numbered; some have three separate identity tags, indicating that three different teams of scientists have recorded them. We hike a mountain path strewn with worn basalt boulders to the hut where we will stay while making our own studies. Meadows cloak the lower slopes wherever trees have been cleared for skiing, but as we climb higher bamboo becomes abundant. At two thousand meters, the bamboo, which goes by the name *Sasa nipponica*, completely dominates open areas and seems to be invading the fir forest. It is a dwarf that grows only about a meter tall, but it compensates for its short stature by spreading aggressively, forming a continuous canopy of evergreen leaves that light cannot penetrate.

Sasa nipponica is one of three species of dwarf bamboo in Japan, and wherever *Sasa* grows in abundance, tree seedlings struggle in vain to compete with it. It not only deprives them of light, but provides cover for seed-eating rodents that thrive in thick ground vegetation where their enemies cannot find them. The few tree seeds that escape the depredations of the mice struggle to survive in the perennial shade. In the northern Japanese island of Hokkaido, *Sasa* forms such an impenetrable thicket that even foresters have difficulty finding their way through it and forestry operations are carried out in winter when the bamboo is submerged beneath a meter of snow. *Sasa* is a demon, but like every character with supernatural strength, from the heroes of the Greek myths to Superman, this demon has a weakness. In fact, *Sasa* has not one Achilles' heel, but two.

The dual key to the success of *Sasa* is a subterranean, creeping stem called a "rhizome" that enables it to spread, and a total dedication of resources to vegetative expansion for twenty or thirty years at a stretch, undiverted by sex or seed production. These two strengths, however, are also where *Sasa*'s weaknesses lie. The rhizome cannot survive in frozen soil, and so *Sasa* only dominates areas where the soil is insulated from the cold air in winter by a sufficient thickness of snow. In exposed places where the snow cover is too thin, *Sasa* does not thrive. Where it does grow, an extraordinary thing happens after two or three decades of uninterrupted expansion. Every *Sasa* shoot in the forest simultaneously flowers, sets seed, and then dies, apparently killed by the effort. Mass flowering of this kind is the rule among Asian bamboos, the record being held by the massive, Chinese species *Phyllostachys bambusoides*, which delays flowering for 120 years.

How a whole hillside of bamboo composed of many unconnected, individual clones synchronizes this collective act of sex and suicide is not known. Synchronized reproduction also occurs in many tree species (usually without fatal results) and is triggered by climatic cues. For example, coniferous trees in the Northern Hemisphere synchronize their seed production this way on a virtually continental scale, but bamboos achieve something even more remarkable. Parts of the same bamboo clone flower and die in synchrony, even when artificially separated and planted on different continents. They can only achieve this synchrony if there is some kind of internal clock that is set ticking when the original plant germinates, establishing the botanical equivalent of Greenwich Mean Time. Clearly, Bamboo Mean Time is transmitted from mother to vegetative daughter shoot without being reset, because when bamboos flower all shoots do so together, irrespective of their individual age.

Only once in a bamboo lifetime are the trees that have the misfortune to share their habitat with *Sasa* freed from its malign influence on their offspring. Only then do tree seedlings have a chance of establishment. But what about the mice? The mice are very probably the evolutionary key to the whole thing. *Sasa* seeds are about the size of a wheat grain and, wheat and bamboo both being members of the grass family, they are just as nutritious. If *Sasa*, or indeed other bamboos, produced regular small crops of nutritious seeds there is little doubt that the whole lot would be polished off by rodents and birds every time. A plant that hides you from your enemies and feeds you into the bargain is a heavenly gift to a mouse. Natural selection is not so generous to mice, however, and in bamboos evolution has found a way around a situation that would otherwise spell extinction for the plants. Instead of producing a few seeds every year as most grasses do, bamboos starve their enemies for decades, and then swamp them with so many seeds that they cannot possibly eat them all. When this happens, rodents and birds leave sufficient bamboo seeds uneaten on the side of the plate to allow bamboos to repopulate. The cost to the bamboo plant of concentrating its reproduction in this way is that it dies in the effort, but not before it has transmitted its genes by way of seeds and pollen, some of which will escape the mice.

The death of bamboos after flowering is an extreme example of the trade-offs that govern the evolution of plants and animals, just as much as they regulate human actions. The truism that you can't have your cake and eat it too, translated into bamboo-ese, reads: "You can't exhaust yourself in seed production and expect to survive." The key point about all trade-offs is that engaging in one activity exacts a cost that must be paid by forgoing some other option because the cake (resource) is of limited size and gets used up. The cost of reproduction is a particular type of trade-off that is highly important in limiting the demonic tendencies of plants and animals, as we shall see when we climb higher up the mountain.

We are now done with *Sasa* and are approaching the Mount Shimagare mountain hut where we shall spend the next few nights. This is where every plant ecologist to visit the mountain in the last two decades or more has made base camp. For the last twenty-two years it has been owned and run by the same hut master, and as we reach the hut he greets my companion Nikki Kachi like an old friend. The hut master puts down his barrow load of fuel wood and shakes us all firmly by the hand, addressing Nikki as *Sensei*, meaning "honorable teacher." In return he is addressed by the formal title *Oyaji-San*, or "Mr. Hut Master." It is as if the two men are merely the latest actors to play two ritual roles. Nikki inherited the responsibility for monitoring research plots established by his predecessor at Tokyo Metropolitan University, Professor Makoto Kimura, and some other person will play *Sensei* when, eventually, Nikki hands on the role of professor. For the moment though, both are vigorous men who show no sign of relinquishing their duties. To the contrary, Nikki relishes his role as field ecologist, even to the point of wearing with pride the scars that come with the hazards of outdoor life. His movements are quick and fluid, almost wired, as though he is constitutionally adapted to the repetitive tasks involved in fieldwork. Measuring tens of thousands of trees in a season is routine. *Oyaji-San* is twenty years Nikki's senior, perhaps in his mid-sixties, and has the face and stature of a jovial Buddha with a goatee beard and gold teeth. His hut is supplied with electricity from a small wind turbine and an array of solar panels, and there is a satellite radio dish on the roof, but inside there are wood paneling, oil lamps, and the permanent reek of woodsmoke.

In the evening we sit around a low table and *Oyaji-San* presides over an evening of sake-drinking and convivial chat. Another guest in the hut is a professor of electronics who sports a flashlamp on a headband that he seems never to switch on or to take off. I speculate whether this is his badge of professional office, like Nikki's scars. It appears that half of *Oyaji-San*'s guests are usually professors of something, somewhere, but he's clearly not daunted by this and seems to have been especially amused by the German professors of philosophy who visited.

A couple of days later, in fine weather, we ascend Mount Shimagare from another direction. There is no view of the peak till about half way up, when the path suddenly breaks out of the forest onto a crag of large boulders. There, laid out before us is a breathtaking sight. The side of the mountain is hanging in front of us, like a green damask cloth suspended in folds from an azure sky. The texture of the cloth ripples with treetops. In the foreground, branches held upward at a thirty-degree angle look like coarse brushstrokes on an artist's canvas. As the view recedes uphill, the texture changes to stippling in two subtly different shades of green. A darker green is the fir *Abies veitchii*, but there is also a more glaucous, grey-green foliage belonging to another fir species, *Abies mariesii*. The two species between them monopolize the whole landscape. Every tree visible from this crag is one fir or the other. You can see from this vantage point that the two species are not perfectly mixed, but patchy in places and intermingled in others. Here are two demons of the subalpine zone that somehow coexist.

On its upper flanks and running across the mountainside near its peak, the damask is slashed and the green canopy gapes open, revealing vertical, grey trunks of fir, like the warp of a threadbare fabric. Some of the threads have come adrift and lie at an angle. Through binoculars, each thread resolves into the weathered, sun-bleached trunk of a dead fir tree. Devoid of all leaves, branches, and bark, these standing dead trees are a telltale sign of the phenomenon we have traveled six thousand miles from England to study. These rents in the fabric of the forest are the breaking crests of fir waves—a freak of nature found only in these high mountains and, nine thousand miles away, in the northern Adirondack Mountains in the United States.

A forty-minute climb from the crag and we reach the lowest of the waves. On the way up we exchange greetings with hikers making the steep descent into the valley: *kon-ichiwa*. Leaving the path, we enter the forest just behind a wave crest and find a grove of slender trunks blotched with lichens as big as the palm of your hand. The trees stand close together, sometimes so close you need to squeeze between them to pass through. The sky has clouded over now, and the grove is illuminated from its downslope side by a diffuse white light that throws the trunks into silhouette, creating the eerie impression that we are in a cage guarded by vastly tall bars.

These trees are as old as any get on Mount Shimagare. With the aid of a tree corer we estimate them to be eighty. The canopy is no more than sixteen meters above the ground—short by normal forest standards, but there are no branches till three-quarters of the way up, and viewed from the ground the taper of the bare trunks creates the illusion that they are taller. The firs' lower branches have been shed, being too far from the sunlight to repay the tree the cost of maintaining them. Every gram of living plant tissue must be supported by water and nutrients and is a drain on the rest of the organism if it cannot support its own energy requirements through photosynthesis and also supply a surplus to support growth. Trees similar to these, but growing by the mountain path, retain their lower branches longer because, in sunlight, they are of value to the plant.

A breeze moves the firs, which oscillate like graceful, pliable wands on a metronome. Though varying little in height, the trunks vary in thickness and, accordingly, each oscillates to its own beat. Thinner trunks also precess, describing an ellipse with the motion of their crowns. Staring up at the canopy for more than a minute or two induces giddiness from the motion of these erratic timekeepers, each moving out of synchrony with the others. The forest floor is an altogether quieter place for the eye, but there are signs of motion here too, though on the scale of generations rather than seconds. The ground underfoot is soft and springy because we are standing on countless generations of rotten tree trunks covered in a thin layer of fir needles and woody debris. Almost as abundant as the standing trees are rotten fir stumps. Reduced to knee height and covered in moss, these are the remains of the generation that died when our eighty-year-olds germinated. They were the eighty-year-olds' parents. Look closely now at these stumps and see that they nurse tiny seedlings growing out of the moss that covers them. These seedlings are the eighty-year-olds' offspring. Seedlings growing on nurse logs stand a much better chance of survival than seedlings germinating directly on the forest floor. What could be better for a seedling than being cosseted in moss on a fibrous, moisture-retaining seedbed raised half a meter above the sedges and herbs that shade the forest floor? Thanks, grandma!

Walk downslope a little way, toward the light that filters through the barred window of this grove, and notice that more and more of the large trees are dead. Not more than fifteen meters farther we reach the edge, where all the large trees are dead. Though still standing, they probably fixed their last molecule of carbon dioxide fifteen or even twenty years ago. The dead trees form the rent in the fabric of the forest canopy that we first saw from the crag halfway up the mountain, and these bare trunks are the exposed threads of its warp. Though the large trees are dead, the forest is not dying, for the seedlings that germinated on grandma's knee fifteen meters back here form a thicket of strapping twelve-year-olds, a meter and a half high and fighting each other for the light. Nothing could be more normal than this handing-on of the baton between generations. It happens in all populations, and in forests it is known as "regeneration."

What is decidedly peculiar in these fir forests on Mount Shimagare is that the generations line up, one behind the other, culminating in the oldest trees that die at the crest of a wave that ushers in the new generation. In normal forests the generations are all mixed up, with a patch of older trees here and a clump of youngsters there where they have colonized a gap left by a fallen tree. Forests that regenerate by filling gaps with seedlings become a mosaic of generations. In a fir wave, by contrast, we have the whole life cycle of a tree laid out for us in chronological order, from wave crest to wave crest. This is a scientific opportunity worth climbing a mountain for!

Let me take you on a walk through the life cycle of a fir tree. This is not a metaphorical walk, but a real journey. By traversing the length of a fir wave we shall travel in time from the germination of a seed to the death of the tree into which it will grow. Of course, it isn't the same tree throughout, but you may safely imagine it to be so. Ecologists call this a "chronosequence," or describe what we shall do as "space-for-time substitution." In fact, we have already started our journey because it begins with those small seedlings growing on grandma's knee. To study this fir wave we have stretched a surveyor's tape marked in meters from the place where the seedlings grow, fifty meters downslope to the next grove of tall trees. Each meter we progress along the tape we estimate the age of all trees and seedlings within a prescribed distance either side of the tape. This kind of sample is called a "transect" and, when used in a fir wave, it allows us to measure the number of individuals belonging to each age class (e.g., one- to three-year-olds, four- to six-year-olds, and so on) in the population. We will learn a lot from this basic census information.

At the start of the transect, there are about thirty seedlings per square meter, a very low number by comparison with counts we have made in fir waves in the Adirondacks, where seedling densities in the year following a big seedfall can reach as high as three thousand per square meter. Densities that high look like a carpet with a thick, green pile. Ageing the seedlings in the Japanese forest gives us a clue as to the cause of this difference. Seedlings and saplings are far too small to age by coring, but luckily there is another, harmless way to find out how old each one is. Firs, like most conifers, produce one new whorl of branches each year. A whorl is a set of four branches arranged symmetrically around the stem. These branches persist for at least a few years, but even after they have been shed, there remains a set of four elliptical scars that can still be visible on the trunk up to thirty years later. By the time most trees are this age, they are large enough to core without damaging them, and estimating age from growth rings confirms that whorl counts are accurate.

Counting the number of whorls down the stem of a young fir seedling is easy, and we rapidly discover that the seedlings at the start of our transect are about five years old. There are no younger seedlings about, so it looks like the mature trees above our heads have not produced seed in the last five years. Such long intervals between seed crops are quite normal in forest trees, particularly up here in the sub-alpine zone. Can seedling densities have dropped a hundredfold from three thousand to a mere thirty per square meter in only five years? Experience from elsewhere, and simple arithmetic, tells us the answer is "Yes, quite easily."

Thirty seedlings remaining out of a possible three thousand means there has been 99 percent mortality in five years. This enormous "wastage" of offspring is the norm in nature, a fact well known to Darwin but perhaps still little appreciated by the world at large. Takashi Kohyama, a plant ecologist at Hokkaido University of whom we shall hear more later, studied fir waves on Mount Shimagare in the late 1970s and estimated that an eighty-year-old *Abies veitchii* tree produces over a quarter of a million seeds during its lifetime. If the population is neither expanding nor contracting in area, an average of only one seed per mature tree itself reaches maturity. One out of a quarter of a million seeds is a mortality rate of 99.9996 percent. This is carnage. Or at least that's what it would be called if animals were killed in such numbers. What is the plant equivalent? Foliage?

Fifteen meters along our transect and all big trees are now dead. We shall find out why these octogenarian firs died when the event is repeated at the end of our transect, but for the moment just consider the consequences for the next generation. The five-year-olds are now twelve. The death of the large trees means that their foliage no longer shades the forest floor and gaps so formed allow light to flood onto the twelve-year-old firs below, making them grow like there is no tomorrow; indeed for many of them there won't be. Some trees at this point are a meter and a half tall, and we can tell from the length of the main stem separating recent whorls that their growth has accelerated, reaching ten or twenty centimeters height growth in a single season. Others are doing poorly and their recent whorls are very close together, showing how little they have grown. These trees are now in the shade of their taller neighbors; they are losing the competition for light and are destined to die. Many of their even smaller neighbors are already dead and the density of living firs is now down to about fifteen per square meter.

Walking along the transect is now becoming difficult. By the twenty-five-meter mark the trees are not only battling each other even more fiercely, but they are bigger than us and are fighting back! We have entered a teenage antechamber of death where the living and the dead stand together, so closely packed that the only way through is to put your head down, shut your eyes, and dive forward with arms outstretched in front of you to part the trees as though penetrating a wave of water rather than a wave of trees. This is the zone called "dog hair" in the Adirondacks; the fir stems are stiff, wiry, and dense, and every one is barbed with dead side branches that whip and scratch. Few ecologists, but even fewer young firs, escape this zone unscathed.

Once clear of the dog hair, another hazard for firs and a new obstacle for us appears. The standing dead trees of the last generation are now so rotten near their bases that storms have brought them crashing down, and the massive logs lie in a tangled wreck on top of the living firs. Grandma was kind to the seedling, but mother's legacy can be awfully cruel. Many young firs are felled or decapitated by falling trunks, though some victims show signs of recovery, signaling that mother's blow may not always be fatal. Successful recovery probably depends on whether a victim's neighbors have also been hit, thus preventing their shading of the damaged tree.

By fifty years of age some trees are large enough to have begun to produce cones; their density is now down to one tree per square meter. Male cones are produced on lower branches, and female ones on one-year-old stems near the tips of high branches. When female cones are ripe they fall apart, scattering scales and seeds to the wind, leaving behind an upright, stalklike structure called a "rachis." Particularly in balsam fir, *Abies balsamea*, in the Adirondacks, but also in the Japanese species here, rachises may persist on the tree for ten years or more. Since new cones appear only on one-year-old stems and each branch grows by one whorl per year, it is possible to work out how old any particular rachis is by counting the number of whorls of branches that lie between its position on a branch and the branch tip. Using this method we can read a tree's reproductive history, retrospectively, just by looking at it.

Imagine, if you could read a person's family history just by looking at them, how informative that would be for the inquisitive student of human affairs! No social scientist is so lucky, but fir trees are the enquiring ecologist's dream. They offer an open book to those who wish to read it. In it, we read a population's statistics of mortality and survival from the way that density reduces as age increases along our transect; we read how old any tree is by counting whorls or tree rings; we read how much a tree has grown in girth in any year from the width of its tree rings, and how much it has grown in height from the length of stem between two whorls. Finally, the rachises recount reproductive history, cone-by-cone, year-by-year. No other organism that cannot speak is so eloquent about itself. What these facts about firs have to tell us we shall learn anon, but we are now approaching the end of the transect and a tale of death and destruction is more pressing.

The trees, or what is left of them, are now nearly eighty years old. Their density has dropped to one or two firs in ten square meters, they have shed all their lower branches, and they are now almost the tallest and almost the oldest trees in the forest. Downslope a short way, slightly taller, slightly older trees are dying. The first sign is a loss of foliage, leaving tree crowns looking scraggy and pathetic, like lopsided diminutive Christmas trees stuck on top of poles fifteen meters high. For a long time no one really understood what caused firs to senesce like this, but the explanation now seems clear. And in that explanation is also revealed the mechanism that drives fir waves themselves. The culprit is wind. Not the gentle breeze that powers the metronome grove of summer, but freezing, incessant winds that blow from the bottom of the mountain all winter. These winds deposit a rime of ice on the foliage of the most exposed tree crowns, which are, of course, the tallest, oldest firs near the ridges on the flanks and summit of Mount Shimagare. In the spring, the rime ice melts and falls, and so does the foliage, shriveled and burned by cold. These trees that decades ago shed their lower branches because they did not benefit the economy of the plant have now lost the foliage at the top, which was all that was left to sustain them. They have burned the candle at both ends, and now starve to death for want of leaves with which to capture sunlight. This is the way the world ends in a fir wave—not with a bang but a whimper.

The eighty-year-olds are now dead. Their crowns are gone, they no longer age, but neither are they fallen. Next winter, downwind and down-transect, with the protection of their neighbors now gone, the next tallest, oldest living trees of the forest are lined up to meet their death in the full force of the freezing wind. And so it goes on, year after year, generation after generation. The crest of the fir wave moves up the mountain driven by the wind. It is a wave of death, but in death there is new life for the generation of seedlings that colonizes the forest floor just ahead of the wave crest. With the appearance of these seedlings the cycle is complete.

As each wave crest recedes, the seedlings left in its wake grow fewer, older, and larger, the survivors ultimately to die when a new wave reaches them. Fir waves move like ripples on a pond. Watch a cork or a boat ride the ripples. The wave moves outward, but the water itself just rises and falls as the wave passes. Fir trees rise and fall as the wave of death passes, but the trees themselves do not move.

How do we know that continual winter winds are responsible for driving fir waves, and not some other mechanism such as disease? The answer is that we do not know for sure, but there is some convincing circumstantial evidence. The observations that waves always move downwind and that they do not develop on more protected slopes strongly suggest that wind must be involved. Tree rings also support the idea that trees starve to death. The wider the rings, the better a tree has been growing, but the rings laid down in an American fir wave clearly showed that growth had almost ceased before death. Most convincing of all is a computer model created by two Japanese ecologists K. Sato and Yoh Iwasa that reproduces fir-wave-like patterns when it is programmed with rules that simulate the real situation.

A computer model is like a virtual forest in which you, rather than nature, make up the rules of how things work. If the virtual forest behaves and looks like the real thing, this is good evidence that you have got the rules of how the real thing works right. The simpler the rules the better. (The more complicated the rules the easier it is to cheat!) A virtual forest needn't only exist on a computer. We could use a chessboard as a model, which is just like what Sato and Iwasa did with their computer version. So, lets imagine a chessboard where the playing pieces are trees. The pieces vary in height but not in color. They are all green. You are not confined to the usual number of pieces, but have a bagfull with which you can fill every square on the chess board. Place pieces on the board, drawing one at a time from the bag without looking till the board is full. There should be no pattern in the heights of the tree pieces. If drawn blind from the bag, they will not be lined up by size as they are in a fir wave.

Now let's play. First, choose a "wind direction" that will last the whole game, simulating the prevailing wind. You could decide for example that the wind blows from left to right. The exact direction doesn't matter so long as it is fixed for the duration of the game. Now the rules. This game is perfect for children because there is only one player, so it's always your turn. Just like a real board game, the rules involve neighbors in adjacent squares. At each go, starting on the upwind side, systematically search the board for any tree that is larger than its immediate upwind neighbor. Remove these pieces. In the jargon used by advanced players of Fir Wave®, these pieces are said to be "rimed." Replace each rimed piece with a seedling taken from the playing bag. These are the smallest pieces in the bag, so be careful not to lose them. When you have checked every square, go back to the beginning and start again. You will find that, as if by magic, each piece on the board has grown a little taller. Replace all the rimed pieces with seedlings and repeat. Do this ten thousand times and then stop. This last rule is why most professionals prefer to play the computer version of Fir Wave®, but the classic version we are playing produces an identical result. As the end of the game approaches you find that trees are lined up by height and that waves of mortality ripple across the board, just like real fir waves. These results depend only on wind direction and the very simple relationships based on the relative heights of next-door neighbors. As well as being a very good way of keeping small children occupied for many hours, this game strongly supports the causative role of directional wind in the production of fir waves. If you doubt it, change the direction of the prevailing wind and play again. It also illustrates a more general point about plants, which is that, being rooted to the spot, it is interactions among immediate neighbors that matter most. If you doubt this conclusion, you can change the rules of the game accordingly and see what happens.

Of course real life is a little more complicated than the model world, and one complication is that in the fir waves on Mount Shimagare there are actually two species of fir, not one. The playing pieces are in two colors after all. Takashi Kohyama, whose work on fir waves has already been mentioned, studied the differences between the two species of fir on Mount Shimagare. He found that *A. veitchii* grows better in conditions with more light, but that *A. mariesii* has the superior growth in shade. Overall, *A. veitchii* is much the more abundant species in fir waves, but the shade tolerance of *A. mariesii* apparently enables it to hang on because there are places in the fir wave where it does better than its competitor. If the balance between these two sub-alpine demons sounds like a precarious one, it is. The two species of *Abies* coexist only in central Honshu. *Abies mariesii* occurs alone in the forests to the north of Mount Shimagare, while *A. veitchii* is the only species to the south. Why the two firs do not share more of their ranges is not known but, being so similar, competition between the species is a likely explanation. The more similar two species, the stronger competition between them is expected to be and the less likely it is that they can coexist. In ecology, difference in structure, habits, or physiology is the key to harmonious coexistence.

Another ecological rule is that similar environments foster similar ecosystems and similar kinds of species, even when those environments are on opposite sides of the globe. Nowhere is this more clearly borne out than when comparing the Adirondacks in upstate New York with this part of the Yatsugatake Mountains in Honshu. Both areas are dominated by species of fir, and in the sub-alpine zone both show wave regeneration. Inhabiting the forest floor, related species of sorrel and dwarf cornel occur in both places too, though there are some floristic differences. Bamboo gives the Japanese mountain flora a distinctly Asian flavor.

Exposed to similar environments, humans behave alike too. Whiteface Mountain in upstate New York, or Fir Wave Central as I prefer to think of it, is a ski resort in winter and an outdoor recreation area in summer. The lowland woods of the area are full of log houses that stand among tall pines. The pines are a different species, but the cabins are there in Yatsugatake too, only bigger. There is skiing on Mount Shimagare in winter and trekking in the mountains in summer. The décor in the restaurant near Mount Shimagare—the "Country Kitchen"—is the same as its counterpart near Whiteface Mountain, right down to the Adirondack recliners fashioned from half-finished timber that adorn its deck, but the Japanese resort is the more upscale. When she toured the state as Senate candidate for

New York in the 2000 elections, Hilary Rodham Clinton was asked why she and Bill didn't vacation in the Adirondacks. Some of her opponents thought they already knew the answer and appeared at her upstate campaign rallies dressed as flies. The woods around Whiteface Mountain swarm in summer with small, black flies with a very painful bite that make life a misery, especially if swatting them means you have to let go of a branch and fall out of your fir tree. Wearing a head net doesn't help much, as somehow the flies always get inside. Mercifully, Mount Shimagare seems free of this scourge.

Whiteface Mountain is truly the world capital of fir waves. It has a paved toll road that takes you nearly to the very top, where a café housed in a granite castle sells what must surely be the best-defended burgers in all the state. The road was built in the 1920s as a memorial to the fallen of World War I and affords spectacular views from the top. On a clear day to the north, you can see the skyscrapers of Montreal, Canada, but it's the view to the south that interests us. In that direction the balsam fir forest beneath the summit is just solid fir waves across the whole mountainside. The forest canopy isn't a fabric with just a few rents in it like that near the summit of Mount Shimagare, but the ragged attire of a beggar with a hundred threadbare tears in his cloak, draped over the mountain.

One demonic trait is to reproduce extravagantly. It was at Whiteface Mountain that we discovered what puts the brakes on reproduction in a Darwinian demon. Theory suggests, and observations verify, that reproduction exacts a cost on the organism. Thankfully the cost of reproduction in humans is not as mercilessly abrupt as it is in bamboos, but think of the effort and material resources it takes to bear and raise children. Personally, I prefer not to calculate the financial cost of raising my two children (what is money for, anyway?), but when my wife went away for three weeks leaving me in sole charge, I lost six kilograms in weight. Every parent knows that reproduction incurs a cost, however you choose to bear or calculate it. Plants are no different, but calculating what it costs a plant to make a seed is not so easy. What better way to investigate this question than in a fir wave populated by loquacious balsam fir?

Using the reproductive history recorded by each fir tree in the location and number of its rachises we estimated how may seed cones each tree in our study had produced in each of the last ten years. We reasoned that the resources a tree uses to make cones and seeds must be taken from those that would otherwise be used in growth. Therefore, a good way to investigate the cost of reproduction in fir trees is to see whether trees with more cones grow less in height. Remember that, like cone production, height growth can be measured retrospectively too: it is recorded in the length of stem between two successive whorls of branches.

Our measurements and rachis counts at Whiteface Mountain showed that, as expected, there is a negative relationship between cone production and height growth. To be precise, producing a single cone reduces the height growth of a fir tree by about two millimeters. "What a bargain! Gimme fifty," you can hear the sex-hungry tree reply. Well, two millimeters may not sound like much, but it adds up. Spend little and produce only a few cones, and the handful of seeds they contain are all likely to be eaten by rodents and birds. It is the same problem that confronts *Sasa*, and the solution is the same: save up and produce lots of seeds at once. Balsam fir at Whiteface Mountain produces a large seed crop every five years or so. When this happens, a tree that produces fifty cones will stunt its growth by ten centimeters. That too may sound like a modest price to pay, but we have to consider what effect stunted growth might have on the future survival of a tree.

Anyone who has walked the length of a fir wave knows that this is a tough place for a tree, a place where the vast majority die before reaching maturity. The law of this particular jungle is "Grow or die!" Taller trees win the competition for light; shorter trees never make it. Imagine what would happen to a teenage mother. To have its growth checked by ten centimeters as a consequence of reproducing prematurely could easily lead to the death of a young tree. By contrast, a tree that delays reproduction and puts its resources entirely into growth until most of the competition is dead will survive to reproduce many times and leave more offspring. Thus, evolution favors delayed reproduction in these populations. This is no doubt why firs at Whiteface Mountain are thirty, or nearly half way through their lives, before they start to bear cones. At Mount Shimagare, where firs live eighty years or so, they do not begin to bear cones until the age of fifty. By coincidence humans live about eighty years too, but human females are at the end of their reproductive period by age fifty, not at the beginning. Emma Darwin bore Charles a child when she was forty-eight, but that was her tenth and last confinement.

The lesson of Shimagare and Whiteface is that even demons capable of dominating the landscape have their weaknesses. *Sasa* can grow aggressively for a period, but eventually it reproduces and dies. It pays the ultimate cost of reproduction. A less extreme example of the same cost is seen in the life history of fir trees that is so beautifully laid out for us in fir waves. Fir trees, like other organisms, must divide available resources between the competing demands of reproduction, growth, and maintenance. Thus, reproduction carries a cost that puts a brake on the runaway success of demons. In the secret weaknesses of demons lie opportunities that other species can exploit. These weaknesses are the key to diversity. There is nowhere better to see this than in tropical forest.

5

The Panama Paradox

There are 1,377 species of native flowering plants in the British Isles. We may have missed a few, but these islands have famously crawled with naturalists for more than three and half centuries and there can be few hiding places left for undiscovered plants. Not so in the tropics. Barro Colorado Island, a single island in Panama with an area of just fifteen square kilometers, which would fit over twenty thousand times into the land area of Britain, is host to 1,212 native flowering plants. This is a fairly typical slice of tropical diversity, but with an extraordinary scientific tale to tell.

Our journey to Barro Colorado Island (BCI) begins at the dockside on Gatun Lake. We reach the dock by taxi from Panama City and await the launch that ferries people and supplies to the island. It is 8:00 A.M., and the day is already beginning to heat up. We should have risen *en la madrugada*—at dawn, as any Spanish conquistador from the hot tablelands of Castille would have done to avoid the heat but, like mad dogs, we'll still be out in the midday sun. The launch arrives and we pile aboard with our backpacks and gear. A few graduate students join us and we're off. There are no tourists because BCI is a nature reserve dedicated to conservation and research. The reserve was created in 1923 when a scientific research station was established on the island, which is why, out of all the tropical forests we might have chosen to visit, we are chugging across Gatun Lake headed for this, the most-studied piece of forest in the tropics.

As we cross the lake we see over to our left an inflatable traveling at high speed parallel to the shore. There are two figures in the boat, one at the tiller of the outboard motor and the other in a combative stance amidships, cradling a heavy machine gun across his chest and sporting a bandanna that streams behind him. This is one of the military patrols that guard the Panama Canal—for that is where we are. Gatun Lake was artificially created by a dam across the river Chagres, forming a vast, lacustrine aqueduct astride the American continental divide. The lake links the Atlantic and Pacific arms of the canal, which connect with it at either end by way of huge locks. The building of the Panama Canal at the start of the twentieth century was the biggest feat of civil engineering in its day. So big that the first attempt, begun two decades earlier, defeated Count Ferdinand de Lesseps, the Frenchman who had successfully constructed the Suez Canal. It bankrupted his grandly named *Compagnie Universelle du Canal Interocéanique*, and he returned to France to face litigation and a decline into insanity. The count had made two fatal errors: against wiser counsel, he had planned an impossibly ambitious sea-level canal that would have required a tunnel through the continental divide, and he underestimated the impact of tropical diseases. Yellow fever and malaria in particular decimated the workforce.

A decade and one Panamanian revolution later, the United States had purchased control of the Canal Zone from the new government of Panama and had begun a second attempt at building a canal. Barro Colorado Island was born of this effort, with Gatun Lake its crib and the United States its parents by adoption. Until 1914 Barro Colorado was just a hilltop, but in that year the rising waters of Gatun Lake made of it an island. Now, the shipping lanes of the Panama Canal carry cruise liners past BCI, but this is as near as tourists get. We shall have a privileged visit. Enjoy the breeze on our short trip, for when we reach the island the air will begin to feel as hot, humid, and heavy as pea soup.

Before long, the island heaves into view and we round a headland into Laboratory Bay. The engines are cut, we tie up at the jetty, and scramble ashore. A steep climb up concrete steps and we are at the old dining hall, a wooden building with a gabled roof that now serves as a museum. Back in the 1920s, this is where James Zetek, the first scientific director of the research station, and his colleagues ate their meals and admired the view of the lake and its farther forested shore. The scientists had originally come to the Canal Zone to help defeat mosquito-borne diseases. Where the French had failed the Americans succeeded because Ronald Ross, working in India, and Walter Reed, in Cuba, had identified the two different species of mosquito responsible for transmitting malaria and yellow fever. Armed with this knowledge, the American administration attacked mosquito breeding sites and were able to control the diseases, though unable to eradicate malaria completely.

The rainfall in this part of the tropics is seasonal, and the alternation between rainy and dry seasons dominates the lives of the animals and plants here almost as much as the alternation between warm and cold seasons does in temperate regions. The diversity of the plants in this forest is not quite as great as it is in the tropical rainforests of the Amazon or Southeast Asia that have no dry season, but it is still extraordinary. If you set yourself the task of learning to identify one new species of tree each day you spent on BCI, you would have to stay a whole year to complete the list. In addition to 365 species of tree, there are 116 shrubs, 265 climbing plants, and 466 herbaceous species resident on this little island. This kind of diversity is a real challenge to the Darwinian-minded ecologist. It is a Panamanian paradox. How do all these species manage to coexist with one another? Where are the demons? What keeps them in check?

In the 1970s an American zoologist named Steve Hubbell began to worry about this problem. He had grown up in Costa Rica and loved the tropics and was familiar with its biological riches at first hand. At the time there were really only two ideas to explain how competing species could coexist, and Steve didn't think either of them could really explain how most tropical trees manage it. The first idea was the niche hypothesis. This requires that competing species in some way share out the resources that are essential to them. Each species must specialize on its own private part of the resource, defined by its niche. The textbook example of this is among Darwin's finches in the Galapagos Islands. Though named for Darwin who collected specimens from the Galapagos in 1835, it was really another English naturalist, David Lack, who discovered a century later that finches have niches. The clue to a finch's niche is its beak. The majority of Darwin's finches eat seeds, and beaks are seed-processing tools. Large beaks are suited to handling tough, large seeds and smaller beaks to handling softer, smaller ones. Thus, the size of a bird's beak reflects its feeding niche. Competition between species forces birds to specialize. Wherever different species of finch come into competition with each other on the same island in the Galapagos archipelago they have evolved differences in beak size. Where those same species of finch occur apart, free from competition with one another, their beaks are similar to one another in size. Competition leads to specialization and specialization permits coexistence.

While the niche hypothesis appears to explain coexistence in many animal communities remarkably well, it is difficult to see how it can work so easily for plants. The problem is that all plants require the same few essential resources and obtain them in a very limited variety of ways. All plants need light, water, nitrogen, phosphorus, potassium, and minor though vital amounts of a few other chemical elements such as iron and magnesium. To be sure, there are some structural differences between plants that are perhaps the equivalent of beak-size differences in birds: there is a tree niche, a shrub niche, a niche for climbers, one for ground-dwelling herbs, and another for epiphytic herbs such as bromeliads and tropical orchids that live on the branches of other plants. However, at BCI there are 365 trees, 116 shrubs, and more. Are there really 365 ways to make a specialized living as a tree? In *On Beyond Zebra*, Theodor Geisel, better known as Dr. Seuss, captured this problem in typically memorable fashion:

> And NUH is the letter I use to spell Nutches
> Who live in small caves, known as Nitches, for hutches.
> These Nutches have troubles, the biggest of which is
> The fact there are many more Nutches than Nitches.
> Each Nutch in a Nitch knows that some other Nutch
> Would like to move into his Nitch very much.
> So each Nutch in a Nitch has to watch that small Nitch
> Or Nutches who haven't got Nitches will snitch.

Plants have the nutch nitch problem bad: much nutch, nix nitch. This problem led two ecologists simultaneously to come up with an alternative hypothesis that now bears their names. In 1970 Dan Janzen, then at the University of Chicago, and Jo Connell at the University of California, independently suggested that the potential demon plants of tropical rainforest are held in check by their natural enemies. Like all the best ideas, Janzen and Connell's enemies hypothesis is elegantly simple. The tropics are rich in species of all kinds, and many of these, perhaps even the majority, are parasites and pathogens: nibblers of leaves or rotters of one species or another. Every tree is plagued by a legion of caterpillars, sap-sucking bugs, insects whose larvae mine for their food between the upper and lower surfaces of leaves, fungi that inhabit the internal tissues or invade the roots. Plants cannot run from these enemies, so they must fight back. It is remarkable that a group of organisms, so singularly alike in how they acquire nutriment, are superlative individualists in their weapons of defence.

The majority of plant species are loaded with chemicals that have biological effects on animals, and no two species have identical arsenals. A significant proportion of our medicines derive from these compounds or are artificial mimics of them: quinine from Amazonian *Chincona* that we use to fight malaria, estrogens used in contraceptive pills from the Mexican vine *Dioscorea floribunda*, the anticancer agent taxol from the Pacific yew (*Taxus brevifolia*), to name but three of a very long list. The variety and individuality of plants' biochemical defenses have forced their natural enemies to evolve specialized means of detoxifying their food. The majority of natural enemies are perforce specialists. Janzen and Connell realized that specialized natural enemies might be the key to tropical tree diversity. Organisms needing special food congregate where they can find it. The particular insects that eat *Quararibea asterolepis*, for example, a common tree on BCI, are headquartered in the canopy of fully grown trees and the same species of caterpillars are found feeding on its saplings nearby. Janzen and Connell suggested that the concentration of natural enemies around full-grown trees would subject their offspring growing nearby to heavy and fatal attack. Only the few offspring arising from seeds that had been dispersed a long way from the parent would escape. If the enemies hypothesis is correct, then demons cannot monopolize tropical forests because natural enemies prevent local concentrations of one species building up.

The enemies hypothesis is one of those ideas that seems obviously right as soon as you hear it. When it was first published, there must have been dozens of ecologists who thought "Damn, why didn't I think of that!" But Steve Hubbell wasn't one of them. His objection was that saplings are *so* concentrated near their parents that even if they suffer much higher mortality there than further away, the end result would still be that a greater number of them would survive near their parents than anywhere else. A little while after Steve published this idea, I asked Dan Janzen what he thought of this objection. "He's plain wrong" was the bold answer. This issue could not be solved by logic alone—hard evidence was needed. Does the Janzen-Connell effect occur and, if it does, is it strong enough or not?

Tropical forests are in some ways ideal for the study of diversity, though in others dauntingly difficult. Quite obviously, the advantage is that there is no shortage of species to study. If you are looking for niche differences or the Janzen-Connell effect, there are plenty of species to choose from. On the other hand, the *majority* of those species are rare, often represented by only one individual per hectare. A hectare is an area of ten thousand square meters, which is a lot of forest to search through to find just one plant when you might need fifty or more for a reasonable sample. Many tropical forests have been censused with one hectare plots, but studies based on samples of this size can say little about rare species because of their very scarcity. A study of tropical forest diversity that leaves out the rare species rather misses the point because the paradox of coexistence is most acute in the case of rarities. To paraphrase, whom a demon would destroy he first makes rare. If we knew how rare species survived competition from common ones, we would have the answer to the whole problem of coexistence because this would tell us what stops species that reach low numbers carrying right on down to extinction. If there is an *advantage* in being rare, as there is for example in escaping natural enemies according to the enemies hypothesis, and we can discover how it works, then we've got the problem licked. In short, the "advantage of rarity" is the crux of coexistence. If we can find such an advantage, it would explain everything.

Steve Hubbell realized that, whatever the answer to the paradox of Panama turned out to be, he just *had* to study the rare as well as the common species. His solution to this practical necessity was both simple and bold: a total census of all the trees in a fifty-hectare plot. Steve resolved to sample half a million square meters of forest! In 1980 Steve Hubbell and Robin Foster, a field ecologist with an intimate knowledge of the tropical trees on BCI, surveyed and marked out a rectangular, fifty-hectare area of forest in the center of BCI. Then, aided by a small army of over fifty field assistants, every stem one centimeter or more in diameter at a height of nearly one and a half meters was identified to species, tagged, and mapped. Steve later described to me his reaction to what they found when this huge logistical exercise was complete: "When we finished the first census I was utterly surprised at the vast number of stems. We had over a quarter of a million stems 1 cm or more in diameter. The other astonishing thing was the number of rare species. We had 150 [species] that together made up less than one percent of all the individuals!" There were over three hundred woody species in all in the plot, so half of the trees and shrubs were rarities.

With the largest dataset for tropical forest trees ever collected at their disposal, Hubbell and Foster started to look for evidence in support of the enemies hypothesis. If the effect was present, then parts of the plot where adult trees of a particular species were abundant ought to contain fewer juveniles (saplings) of that species than places where adults were scarce or absent. Out of forty-eight common species tested for this pattern, it was present to some degree in just under half. However, in all species bar one, the effect was too weak to limit abundance effectively. The exception was the most abundant canopy tree, *Trichilia tuberculata*, which did show evidence of a Janzen-Connell effect that would be strong enough to limit its numbers. Nonetheless, on average every eighth tree found in the plot belongs to this demon, so its numbers are controlled only at a very high level of abundance. The natural enemy that limits *Trichilia* appears to be a chronic fungal disease present in the heartwood of adult trees, which infects nearby seedlings through their roots.

The results of the first census of the fifty-hectare plot at BCI did not suggest that the enemies hypothesis was of *general* importance. Neither could niches offer a general answer, and so from two possible explanations for tropical-tree coexistence we were down to none. Niche hypothesis nil, enemies hypothesis nil. Out of two nullified hypotheses came Hubbell's alternative—a null model, which is a mathematical description of what the world should look like if everything happens at random. It is the world according to the "dice man" of Luke Rhinehart's novel (of the same name), whose every important decision was made by the roll of a die. Could chance alone maintain the diversity of tropical forest? In 1986 Steve Hubbell and Robin Foster suggested it could. They said we were wrong to be looking for differences between plant species to explain their ability to coexist. The whole point, Hubbell and Foster maintained, was that tropical trees coexisted precisely *because* they were so similar. Most people find all but a few tropical forest trees so difficult to tell apart that this seemed a very plausible argument. All the more so because Robin Foster is someone who famously really does know one tree species on BCI from another. If *he* thought many tropical trees behave alike, then it might actually be so.

Hubbell and Foster's argument was similar to one made a couple of years earlier by Goren Ågren and Toby Fagerström, then at Lund University in Sweden. I remember reading their paper soon after it was published, when I was myself still puzzling over how chalk grassland plants coexist (see chapter 6), and finding their argument an equal puzzle. The suggestion that plants coexist *because* they are so similar seemed totally upside down to me. In fact, it was not so much upside down as revolutionary. The point made by both pairs of authors was that, when species are very similar in their competitive abilities and there is some randomness in the fates of individuals, it can take so long for one species to competitively displace another that on any reasonable time scale they effectively coexist. If the numbers of most species weren't controlled by competition or Janzen-Connell effects, then they could just be drifting randomly.

All three hypotheses we have discussed depend on assumptions regarding the similarity of plant species. However, those assumptions are very different from one another. Under the null model, coexistence depends on species being effectively identical to one another in competitive ability. By contrast, niches depend on dissimilarity, and the enemies hypothesis assumes that species, through the effects of specialized natural enemies, respond selectively to their own density. One might have thought that a contest between such diametrically opposed hypotheses would be very quickly resolved, but it has taken twenty years for the evidence to be gathered, digested, and interpreted. The reason it has taken so long is that static patterns of distribution can be very misleading and the evidence of changes with time was needed. The reason we now have that evidence is the genius of the fifty-hectare plot at BCI: the fifth census was completed in the year 2000. And with more data have also come more ideas to explain plant coexistence. I think its high time we visited the famous plot, don't you?

First, we must check in at the administration building. The research station is run by the Smithsonian Institution and its facilities on BCI are superb. Enjoy the air-conditioned comfort of the cafeteria and bathe in its cool air. The very thought of returning here later will revive you when (quite soon) the heat of BCI becomes almost insupportable if you are a temperate-type like me. The sleeping accommodation on BCI is full with research visitors from all over the world, so later we are going to have to share the basement of a wooden shack—

a survival from the early days. At least it has windows, but check the mosquito screens for holes. Lucky we brought our own mosquito nets. Bags are left safely in the shack, and with water bottles and notebooks we are off. We follow a path uphill and, as we enter the forest and leave the intense glare of the noonday sun, the temperature mercifully drops. The plot is in the center of the island, at the top of the hill, so if you are not reasonably fit it could be tough, but it isn't far. I'll save my sympathy for the BBC camera crew we have brought with us who are lugging heavy tripods, cameras, batteries, sound recording equipment, and even lights. We are making a television program and instructional CD titled *The Conundrum of Coexistence* for our ecology course at Open University.

The moment you enter the forest it is the sounds, rather than the sights, that capture you. There is an intense, almost electrical buzzing and whining in the air. It is a chorus of cicadas. Their large, fat larvae inhabit the soil and feed on tree roots—another natural enemy the plants have to contend with. At maturity the winged adults emerge for a brief life of whine, sex, and song. A rarer sound you can sometimes hear if you stop and listen are mantled howler monkeys. These are also herbivores, but they eat fruit as well as leaves and so provide some benefit to plants by dispersing seeds. Their call is an eerie bark, like a howling dog. The "cheep-cheep" of small birds can also be heard, though they are not as abundant as once they were. The isolation of BCI from the surrounding forest caused by the raising of Gatun Lake has taken a progressive toll on its birds. Sixty-five species have disappeared and some others are on the brink of extinction. In 1970, there were on the island an estimated five hundred individuals of a finch called the slate-colored grosbeak, but a survey made twenty-five years later found only two pairs.

The bird extinctions on BCI offer a warning of what must now be happening in forest remnants throughout the tropics where logging, at best, is producing a fragmented landscape. In the particular case of BCI, its isolation has produced some beneficial effects too. Because it is accessible only by boat, and because the island is intensively patrolled by guards employed by the Smithsonian, poachers who shoot large animals in the mainland forest have been kept away. Coatis, a racoonlike animal, and agoutis, which are large rodents, are four times more abundant on BCI than in forest on the nearby Gigante peninsula. Red-tailed squirrels are twice as abundant on BCI. All these animals eat tree seeds that they collect from where they fall. A fraction of the seeds that they collect are removed a distance from the parent tree and then dropped, or hidden and then not retrieved. Thus, these seed-eaters are also seed-dispersers. Trees use these mammals, and larger seed-eating birds too, as transport for their offspring. The fare for this service is extortionate, since the carrier eats most of the passengers, but from the perspective of the parent tree there is a net gain because there is no other way to colonize new sites.

Let's pause a moment here to consider the evolution of dispersal more generally, for it contains another fascinating paradox. The paradox is easily stated: all organisms disperse a significant fraction of their offspring, and yet most of the offspring die. Wouldn't it be better to keep most offspring at home? The answer is nearly always "no," and in 1977 Bill Hamilton and Bob May, evolutionary biologists then at Imperial College, London, came up with the reason why. Imagine an annual plant that disperses none of its seeds. Each year all such plants produce seeds and then die. Seedlings in the new generation all have to compete for the home site, but there is only room for one of them to replace the parent. The more seedlings there are, therefore, the more die in competition with their siblings. Now, let's see what happens if a genetic mutation occurs, causing the plant carrying it to disperse a fraction of its offspring. One of this plant's seeds is certain to recolonize the home site, and *in addition* each of the dispersed seeds has some small but significant opportunity to establish in another site. Nondispersers, by definition, can never do better than to recapture their home site. Plants carrying the gene for dispersal can always do better than this and consequently their numbers increase. That's all it takes to make dispersal a better strategy than nondispersal and to explain why all organisms do it. Hamilton and May discovered that even when the probability of dispersed offspring surviving is nearly zero, the mathematics of the process predict that types that disperse half their offspring will do better than those dispersing fewer. This counterintuitive result has now stood the test of time and is one of the best examples of how evolutionary theory can illuminate the mysterious ways of nature.

Theory tells us that dispersal is more important than we might have imagined. What happens when there are no dispersers? Joe Wright, an ecologist who has worked for the Smithsonian in Panama for many years, investigated the effect of poaching of the mammals that disperse seeds of two palms that are found on BCI and the mainland, including the Gigante peninsula. On BCI, where agoutis, coatis, and other mammals are protected, more palm seeds were eaten by these animals, more were dispersed, and fewer seeds were destroyed by seed-eating beetles. These beetles are among the natural enemies that congregate around adult palms. Elsewhere, where poachers are active, palm seedling densities were higher around fruiting adults, though how many would survive to adulthood is not known.

As we continue our climb uphill, notice the density of stems all around. There are occasional very large trees, stilt palms whose slender stems are supported by prop-roots near the base and some other eye-catching plants, but it is the density of stems of all sizes that really impresses. Everywhere you look there are more and more plants! The slender stems of saplings stand stiff and erect, but there are also lots of woody climbers, or lianas, in this forest whose twisting trunks are festooned in loops from the branches of trees or that snake their way upward toward the canopy and the light. In northern temperate forests climbers such as ivy, *Clematis*, and honeysuckle are mainly decorative, but in the tropics they are a major structural feature of the forest.

There are 171 species of liana alone on BCI, plus many other nonwoody climbers. The Swiss Cheese plant, or *Monstera deliciosa*, which you may have somewhere in your home, is a tropical liana. Climbers have a fascinating biology and Darwin wrote a whole book about them. Whereas most plants are phototropic and grow toward the light from the moment of germination, *Monstera gigantea* seedlings are skototropic and seek darkness. Why they behave in such a contrary manner is obvious if you think about it. They begin life on the forest floor a long way from the light, which can only be reached by first finding a support to grow up. Trees cast shade, and that shade is a cue to the whereabouts of a tree trunk. It is how you would find a tree trunk if you were wearing spectacles with frosted lenses that allowed through light but no image. Plants can sense the amount and spectral quality of light, but cannot see.

There is little color in the understory, so a moving line of small, bright yellow objects bumping across our path is an arresting sight. On closer examination each tiny yellow flag is being carried in the jaws of an ant. Tens of thousands of them form a trail down a nearby tree trunk and along the forest floor for many meters before disappearing into their nest. There is two-way traffic along the trail, cargo being transported by ants traveling toward the nest and ants with empty jaws returning along the same route to collect a fresh load. These are leafcutter ants, and the yellow flags these particular ones are carrying at the moment are pieces of petal that they have snipped from the flowers of a leguminous tree that is flowering somewhere above us, out of sight from the ground.

Leafcutter ants are exceptional among tropical herbivorous insects in harvesting leaves from an extremely wide range of species of plant that collectively probably contain most of the biological toxins known to humankind, and a few more besides. Eating this poisonous cocktail is beyond any animal, but the ants don't consume the leaves directly. Instead, they maintain a fungus garden inside the nest to which they feed the leaf pieces. The fungus digests them and produces small growths on which the ants feed. In return, the ants feed and weed the fungal colony and even propagate it by taking a cutting to each new nest. It's a farm: toxic, indigestible material goes in and nutritious, palatable food comes out. The fungus and the ant are entirely dependent on one another in a relationship that is millions of years old. In fact, it has been estimated from phylogenetic studies that one group of leafcutter ant species has been propagating the same fungal clone for 23 million years!

Leafcutter ants play another role in the life of plants, for they are not only herbivores but also seed dispersers. Seeds of certain plant species have a small oil-rich appendage called an "elaiosome." Ants collect these seeds, transport them to the nest where the elaiosome is removed, and then dump the seed on their refuse heap outside the nest. Seeds stripped of the elaiosome are still viable and may germinate. Larger worker ants in a colony keep trails free from obstacles by removing any leaf litter that lies in their path. Studies on BCI have found that ants from a single nest of the leafcutter *Atta columbica* may build trails with a combined length of 2.7 kilometers in a year. Although no individual trail would probably be more than a hundred meters in length, 2.7 kilometers of trails imply that considerable amounts of seed movement can take place.

We have now reached the fifty-hectare plot. Apart from inconspicuous tree tags and marker posts, there is little to indicate we are now at the hub of Hubbell's grand design. This piece of forest is reassuringly like all the rest, to my eyes at least. One could imagine how easily the forest might be changed by the numerous visitors carrying tape measures and clip boards. In fact, the first fifteen years of census information revealed that the forest *is* changing, but from a much more sinister cause. Near the center of the plot is a small swamp where a number of moisture-requiring species such as the wild oil palm are found. There is a niche, or microhabitat, here for species that are sensitive to drought. Rick Condit, an ecologist at the Smithsonian who has crunched the massive numbers coming out of the fifty-hectare plot, has discovered that nearly all the tree and shrub species that occupy this microhabitat are declining in abundance because the climate of Panama is becoming drier and the dry season is becoming longer. The probable cause is global warming. If current trends continue, significant numbers of species may disappear.

Walking further through the plot we see a piece of red flagging tape hanging from a low branch with the word *Paraponera!* written on it in bold black letters. Is this some kind of warning in Spanish (my Spanish isn't too good)? Almost right. It's actually a warning in Latin. *Paraponera clavata* is a giant ant whose workers can be up to twenty-two millimeters (nearly an inch) long—the largest in Central America. It has a vicious, very painful sting and a colony has a nest at the base of the flagged tree. There are over three hundred nests of this unpleasant creature in the plot. A study of these colonies has found that about half of them contain a tailless whip scorpion called *Phrynus gervaisii*, related to tarantulas. Curiously, *Paraponera* tolerates the whip scorpion, and colonies harboring one have a much lower mortality rate than those without. Perhaps the reason for this lower mortality is that although ants the size of *Paraponera* have few enemies, they are parasitized by scuttle flies that lay their eggs inside the ants' bodies. Whip scorpions prey on insects and may attack scuttle flies. The principle is a simple one: my enemy's enemy is my friend. The tropical forest is full of unexplored, complex ecological interactions of this kind.

Nearby in the forest stands a square frame mounted on four legs supporting a baglike net. The net is not designed to catch animals but plants, or tree seeds and fruit to be more precise. Joe Wright and colleagues have been constantly monitoring two hundred of these seed traps for ten years, identifying and counting all the species that land in each one, and they have discovered something very interesting indeed. During a decade of observation each trap caught on average only 12 percent of the tree species found in the fifty-hectare plot, but each trap caught a somewhat different 12 percent. These findings mean that no species has been able to spread its seeds throughout the forest, and that different species are likely to colonize different places. Over approximately the same period Steve Hubbell, Robin Foster, and their colleagues recorded which tree species colonized twelve hundred light gaps that appeared where trees had fallen in the plot. Their data showed a very similar pattern: individual species colonized on average only between 1.4 and 6.2 percent of gaps.

These findings lend support to a fourth hypothesis to explain coexistence. The problem of how weaker species survive in competition with stronger ones can be solved if the weaker species can escape to colonization sites in the forest (that is, to light gaps) that the stronger competitors have difficulty reaching. What a species lacks in competitive ability it can compensate for by good powers of dispersal. It is not true in nature that "The race is not to the swift, nor the battle to the strong," but it may be true that being swift and being strong are *alternative* and equally viable strategies for survival. This idea is known as the "dispersal limitation" hypothesis, because highly competitive species are usually limited by poor powers of dispersal. Mathematical models of dispersal limitation have shown that only two things are required for it to work as a mechanism of coexistence. First, there must be a trade-off (negative correlation) between species' prowess in competition and dispersal. Second, there must be a lottery for colonization sites. Chance plays a role in determining which species colonize which light gaps. A tree's lottery tickets are its seeds. Good competitors have rather few tickets in the lottery, but every one's a winner. If they get to a gap, its theirs! Poor competitors have many more tickets. Some of these land them in the same gaps as good competitors against whom they always lose, but they also get to gaps that good competitors have not reached and they win these by default.

The discoveries made from a decade studying dispersal and colonization in the fifty-hectare plot strongly suggest that there is indeed a lottery for light gaps. The evidence that the required trade-off between competitive ability and dispersal also exists is less direct, but from what we know about plants it would be very surprising if there were no such trade-off. A key characteristic of plants that are good competitors as seedlings is that they are well endowed by mother with big seeds containing lots of food. Though well provisioned, large seeds have two inherent handicaps that limit the number of gaps that the species can colonize. First, if a plant produces large seeds they are usually few in number simply because the total resources that a plant can devote to seed production are limited. A plant must allocate its resources between either a few big seeds or many small ones. The second factor that handicaps the colonization success of large seeds is size itself. Small seeds travel farther.

Is dispersal limitation then the solution to the Panama paradox? Well, likely yes, but probably only in part. Further discoveries now support the enemies hypothesis more strongly than did Hubbell and Foster's initial census of the fifty-hectare plot. A very important clue has come from comparing the species composition of the seeds falling into Joe Wright's two hundred seed traps with the composition of seedling populations growing near the traps. This comparison tells us what happens between seedfall and seedling establishment. What one might expect is that species that are rare in the seed traps become even rarer by the seedling stage because their small numbers make them more vulnerable to chance mortality. Amazingly, quite the reverse happens and rarer species survive better. In the interval between seeds landing at a site and seedlings becoming established there is selective mortality that takes a heavier toll of common species and a lesser toll of the rare. This effect is that elusive advantage of rarity that is all-important to explaining coexistence. But why are rare seedlings at an advantage? Could it be natural enemies at work? Indeed it could.

As the data from repeat censuses became available over longer and longer periods it was possible to estimate rates of mortality for each species in the plot more accurately. The surprise was that mortality rates of the smallest saplings, those about one centimeter in diameter, were very low. Steve Hubbell told me: "The plants that we thought were young, perhaps two to five years old, turned out to be twenty to forty years old, and that was a serious problem for the way that we had hoped to test the enemies hypothesis because the adult trees die at [a rate of] about 1–2 percent per year. So, in twenty years 20 percent of the parents that produced these saplings may be dead!" If you cannot reliably measure the distance between a young tree and its parent, how can you test whether saplings survive better farther away from mother? Fortunately, all that was needed to solve this problem was a slight rethink of the hypothesis. Young trees are vulnerable to infections by pathogens and infestations by insects that may come from *any* nearby member of the same species, whether a parent or not. There is really no need to differentiate between parents and nonparents. All that is needed is an estimate of the total abundance of a species in the neighborhood of a sapling. If the mortality rate of youngsters is greater at high density than at low, then this will achieve the same effect as the Janzen-Connell mechanism as it was originally conceived.

The data on mortality and the appearance of new individuals up to and including the 1995 census were re-analyzed using this approach, and hey presto! Whereas before there was very little evidence of Janzen-Connell effects, now sixty-seven out of eighty-four species tested showed an effect that favored a species where it was rare. We cannot be sure that these effects were caused by natural enemies in all cases, but some detailed investigations of a few selected species have proved this to be so. For example, Joe Wright found that seeds of a palm called *Scheelea zonensis* had to be at least a hundred meters from their source tree to escape attack by a seed-eating beetle. Greg Gilbert, a graduate student working with Steve and Robin, found that a canker disease affecting the large tree *Ocotea whitei* was more common and caused greater mortality among seedlings near adults of the same species than elsewhere. However, even six years after germination, seedlings of this species were still abundant beneath parents because starting densities beneath them were so high. This was precisely Steve Hubbell's original objection to the enemies hypothesis: the processes proposed may happen, but are they powerful enough? What has changed since then is that now there is such a lot of additional evidence of rare species advantage in the fifty-hectare plot that Steve and others believe that natural enemies seem the most plausible explanation. Only time will tell for sure.

Its nearly time to leave BCI. What have we learned from the fifty-hectare plot in its first twenty years? Albert Einstein once famously remarked that he didn't believe that God played dice. Neither do demons. Chance plays little role in controlling the dominating proclivities of trees species over one another. Natural enemies do that job. The null model of a world in which all is left to chance is largely wrong. However, there is still a small part for chance in another role, as a cog in the mechanism of dispersal limitation. How important dispersal limitation is to coexistence in tropical forest we still do not know. It may depend on whether the expected trade-off between competitive and dispersal abilities is as strong as expected. Finally, niche separation, never a strong contender as an explanation for coexistence in plants, plays a small but significant part in reducing competition between plants in tropical forest. There are other plant communities where, surprisingly, niche separation may matter more. We shall visit these places and their plants in the next chapter.

6

Nix Nitch

It is five in the morning in early summer. We are up with the skylark, and in lark habitat. With a grinding of gears, the municipal bus, which has labored uphill from the town of Brighton to the top of the downs, has deposited us in the chill, bright air. We are standing on the suburban rim, where the spreading town has swept inland from the sea and crested the very edge of the downs, depositing a tide-mark of houses. Beneath us to the right is the English Channel, glistening in the early morning sun. To our left are fields of knee-high wheat, and slicing through their midst is a rutted track, grey-white with chalk and littered with flints. This is an old drovers' road, an indirect clue as to why we have come here. More will be revealed if we follow the track.

As the last houses vanish from view, the open, treeless landscape spreads before us to the near horizon. The horizon is almost unnaturally close, because the track is climbing the back of a gently arching hill that dips down again and out of sight not far ahead. Suddenly, a skylark commences its song, rising from his nest in the wheat. He will rise in spiral flight twenty, forty, even a hundred meters above his territory, singing as he climbs. This limpid, liquid sound once sparked an artistic chain reaction. It inspired the poet George Meredith to write *The Lark Ascending:*

> He rises and begins to round,
> He drops the silver chain of sound,
> Of many links without a break,
> In chirrup, whistle, slur and shake,
> All intervolved and spreading wide,
> Like water-dimples down a tide
> Where ripple ripple overcurls
> And eddy into eddy whirls;
> A press of hurried notes that run

So fleet they scarce are more than one,
Yet changeingly the trills repeat
And linger ringing while they fleet . . . .

The poem in turn inspired the composer Ralph Vaughan Williams to capture the skylark's ascent in music for the violin: a sublime piece that evokes the English countryside as we would wish to imagine it, bucolic and without the cars that nowadays overrun it. In the 1880s Vaughan Williams attended school about a mile from this spot, and perhaps he heard an ancestor of our very bird climbing into the vaulted blue. Back then, these fields were covered in close-cropped turf and flocks of sheep roamed the downs. The drover's road we are following was the route by which, for centuries, the sheep were taken to market. Along the sides of the track there are remnants still of the downland flora that fed the sheep: lesser and greater knapweeds, lesser burnet saxifrage, field scabious, oxeye daisy, wild carrot and wild parsnip, restharrow and cocksfoot grass. The downs either side of the track were plowed half a century ago and the turf is long gone. Over the brow of the hill, we are now descending toward our goal—Castle Hill National Nature Reserve. Ahead, we can see the gorse bushes that cap the highest parts of the reserve with dark green foliage. "Kissing is in season when the gorse is in bloom," goes the saying. There are *always* brilliant yellow flowers on the gorse.

The gate to the nature reserve lies at the head of a secluded valley. The valley sides are too steep to plow, so downland grassland still survives on its slopes. Flowers and blue butterflies are perhaps how most people who know chalk grassland think of it: orchids such as common spotted and early purple, bee orchids that trick their pollinators into visiting them with the false promise of sex, the yellow flowers of birds-foot trefoil on which the blue butterflies lay their eggs, the nodding, yellow heads of cowslips, tall oxeye daisies, tiny violets and many others. But don't expect to see a carpet of color or to find all of these flowers blooming in one visit. There is, however, an ever-present spectacle, painted in subtle shades of green, and easily missed if you have not come face-to-face with it.

What is special about this kind of grassland is the sheer density of its botanical riches. A square foot of turf, or just over one-tenth of a square meter, can contain thirty different species of flowering plants, all of them shorter than ankle height and some real miniatures. The delights of this botanical mosaic are in its details that can be appreciated only at ground level. Aficionados of chalk grassland spend a lot of time on their stomachs and get sore elbows. At this scale, eyeball-to-eyeball with the vegetation, it is the various leaves found so closely intermingled that provide the spectacle, not the flowers.

Peering into the vegetation from ground level you can see that the matrix in which the mosaic is set is provided by a grass with fine hairs placed regularly along the edges of its leaf blades, like the teeth of a comb. This is upright brome. Another ubiquitous grass is sheep's fescue, a plant with needle-fine leaves that tends to form small clumps or mats. The other plants form a collection of leaf-shape motifs that recur in the mosaic. There is the feathery-leaved motif that appears in the leaves of wild carrot, yarrow, dropwort, and lesser burnet saxifrage; there is the cloverleaf motif found in several variations among birdsfoot trefoil, medic, and white clover; there are the plants that scramble around like the delicate fairy flax, wild thyme, and ladies' bedstraw; and there are sedges whose grooved leaves are superficially grasslike, but are arranged, not in opposite pairs as in grasses, but in threes.

If tropical forest is a work painted in a thousand species on nature's grandest canvas, then chalk grassland is a fine-grained mosaic worked in miniature from a restricted palette. A hectare (ten thousand square meters) of tropical forest contains many times more species than a hectare of chalk grassland, but at a smaller scale the tables of diversity are turned. There are more species packed into a square meter of chalk grassland than into a square meter of tropical forest. How do thirty or forty species of grassland plants—all requiring light, water, and the same soil nutrients—manage to coexist in so confined an area without distinct niches? It is the Panama paradox all over again, this time in miniature, or as Dr. Seuss would have it: nix nitch.

Plant diversity, whether in forest or grassland, is a puzzle. The paradox, that somehow similar species compete with one another and yet coexist, is the same for chalk grassland as for tropical forest and the possible solutions to the paradox are, at least in theory, the same. Tropical forest may capture the imagination, but species-rich grasslands offers us greater opportunities to find solutions to the paradox. Things happen quicker in grassland because the plants in these communities are shorter-lived than trees and manipulating them is easier.

Fundamentally, anything that causes rare species to have an advantage over common ones or that reduces competition all round can explain coexistence. The first question is: just how much do chalk grassland plants compete with each other? After all, if the answer is "not much," perhaps there is no paradox. Then, the explanation of how chalk grassland plants coexist would simply be that they don't compete. We know that forest trees compete fiercely with one another and that this happens because one tree can overtop and shade its neighbors (chapter 4). Grazed chalk grassland has a canopy only fifty millimeters tall. Is this sufficient for one plant to overtop and exclude another? The answer to this question is easily found if grazing sheep and rabbits are fenced out, allowing the vegetation to grow taller. Wherever this happens the plant community invariably becomes dominated by tall grasses and other species begin to disappear. The longer that chalk grassland is left without grazing, the more species are lost.

The grasses in the community are fettered demons, only kept in check by herbivores. This is especially true of one particular species, tor grass (*Brachypodium pinnatum*), which has broad leaves with a characteristic yellow cast. Tor grass is unpalatable to grazing animals. Sheep will eat it when it is in small amounts mixed in with more palatable species, but if grazing pressure is relaxed for a while, tor grass spreads to form clumps that sheep then totally avoid. Once clumps of tor grass have formed, there is no stopping its advance as, protected from grazing by tough, unpalatable leaves, this demon grass advances on the surrounding area that sheep have obligingly grazed into submission. There is one slope at Castle Hill where tor grass has spread and excluded nearly all other species.

The experiment of removing grazers tells us that the competitive influence of the tallest species on shorter ones is held in check by herbivory. Grazers fell the demon grasses. There is a parallel here with the effect of natural enemies in limiting the abundance of potential demon trees in tropical forest (chapter 5). In both cases the most abundant plants are the ones to suffer most from the attentions of herbivores, and this creates space for other species. Do grazing and limited competition for light provide the answer to the paradox of coexistence? Some other evidence suggests that it is not the whole story.

We must not forget that as well as competing above ground for light, plants compete below ground for nutrients and water. The soil in chalk grassland habitats is usually quite thin, and solid chalk may lie only thirty centimeters or less beneath the surface. Species diversity varies quite markedly with the depth of the soil. Would you expect more species on deeper soils or more on shallower ones? In fact the most species-rich chalk grassland occurs on the shallower soils. The reason is that shallow soils contain few nutrients and this limits the growth of demons even more than grazing does. Grasses can grow fast and tall, but they need a plentiful supply of nitrogen and water to realize this potential. Without these supplies, tor grass and upright brome are reduced to seven-gram weaklings. However, only a minority of chalk grassland grows on soil so thin that the grasses are completely subdued. In most places there is still competition between plants, even when there are grazing animals about. If competition does not occur above ground, it probably occurs beneath.

Throughout the 1980s and into the 1990s one could visit Castle Hill in spring and be sure of finding a tall, bearded figure prostrate on the turf, as if paying homage to Flora. Peter Grubb, Professor of Plant Ecology at Cambridge University, knows as much about plants and vegetation as any person alive. His signature beard, of Darwinian proportions, sprang into growth in New Guinea when his electric razor broke, and he has sported it since in every corner of the globe. What brought Peter back to Castle Hill every year for a decade was his wanting to learn how the short-lived species that are found in the chalk grassland flora manage to coexist with the perennial species. How do populations of plants such as dwarf centaury, autumn gentian, and yellow-wort that complete their life cycle in under two years manage to persist among their perennial competitors?

By making painstaking counts, year after year, of thousands of plants growing along transects at Castle Hill, Peter discovered what he describes as "drifting clouds of abundance" among the short-lived species. He found that a species might be abundant in one part of his transect in one year, but rare or absent in that same spot in another year and abundant somewhere else. Just as there are gaps in the forest canopy, there are also gaps on a Lilliputian scale in the canopy of chalk grassland created by grazing animals. Gaps provide colonists with a refuge from the roots of competitors as well as from their leaves. It seems that short-lived plants track these evanescent windows of opportunity and coexist with perennials by being able to complete their life cycle before perennials grow back into them. Thus, gaps provide short-lived species with a distinct, short-lived niche. How the different short-lived species coexist with each other *within* their joint niche is another matter and less clear. Peter's counts revealed that seedlings sharing a gap *do* compete with one another, and so this question does require an answer. The dispersal limitation hypothesis (chapter 5) offers one possible solution.

Though perennials may live for decades rather than just a year or two, they too must eventually replace themselves from seed if the species are to persist in the community. Like short-lived plants, seedling perennials may also require gaps for successful establishment, and for the same reason: as a refuge from competition. Peter Grubb suggested that different species may require different kinds of gap for successful regeneration, or what he described as different "regeneration niches." Could regeneration niches explain how chalk grassland perennials coexist? It occurred to me that regeneration niches might be rather subtly different from one another, and that a search for them would have to be conducted at a suitably small scale. I decided to have a go.

The problem of coexistence is at its most acute between related species that share the same habitat, precisely because related species tend to be similar to each other. Darwin made this very observation in the chapter of *The Origin* titled "The Struggle for Existence": "As species of the same genus have usually, though by no means invariably, some similarity in habits and constitution, and always in structure, the struggle will generally be more severe between species of the same genus, when they come into competition with each other, than between species of distinct genera."

Accordingly, I chose to study three pairs of species that all grow at Castle Hill—two knapweeds in the genus *Centaurea*, two plantains in the genus *Plantago*, and two hawkbits in the genus *Leontodon*. With the help of a research student I planted seeds of each of the species one at a time using forceps into marked locations in plots at Castle Hill. The sowing location of each seed was given a score from 0 to 4 depending on how much of the one-by-one-centimeter square of ground around it was covered by leaves: 0 for no leaves, 1 for a quarter covered, and so on up to 4 for complete leaf cover. After that, we painstakingly recorded which seeds came up and how long each seedling survived. After six months, we analyzed the data to see whether species in the same genus "preferred" different regeneration conditions. The result is easily stated: "nix regeneration nitch." There were no major differences.

Though the result of our regeneration niche experiment was clear, its interpretation was not so simple. Should one conclude that, because a single attempt to find regeneration niches failed to reveal differences between species, regeneration niches don't exist? Clearly not, because we might just have made the wrong kind of measurements or, like the drunk looking for his lost keys, we may have been looking in the wrong place. The hapless fellow was wandering around under a street lamp when a passer-by offered to help and asked him where he'd dropped the keys.

"Over there," he replied, pointing into the darkness.
"Why are you searching over *here*, then?" asked the mystified stranger.
"Well, I can't see in the dark, can I?" replied the drunk.

Were we looking in the wrong place for regeneration niches? Not only is there no way to answer this question, but it is quite possible that there was nothing to find. It's as if the passer-by had asked, "What have you lost?" and we were forced to reply "Maybe nothing, but I'll know if I find it." We'll only know whether niches exist when we find them.

Fruitless years of searching for plant niches, and the dawning possibility that other mechanisms of the kind discussed in chapter 5 might explain plant coexistence, eventually led me to the conclusion that there were none to be found. Then, one day in 1997, a plant ecologist by the name of David Gowing, who was working at the time at Cranfield University, came to the Open University to give a seminar about his work on meadows. David had been studying how the flora of these species-rich damp grasslands is affected by hydrology. Meadows are rich in wildflowers and were once common in England, but are now a rare habitat that David's research aimed to understand and protect. He talked about his work at North Meadow in Cricklade, Gloucestershire, which is famous as one of the best sites in Europe for the beautiful snake's-head fritillary (*Fritillaria meleagris*).

Cricklade is a flood-meadow by the banks of the Thames and David had been studying how changes in the drainage of the meadow might affect its flora. He and his colleagues at Cranfield had made a map of the hydrology of the meadow showing in immense detail exactly how long any particular spot was subject to flooding in winter and spring and how long drought affected different areas in summer. Prolonged flooding is a problem for plants because waterlogged soil becomes deprived of oxygen. The roots of plants that are not adapted to such conditions die from lack of oxygen and rot. (Incidentally, overwatering and root rot are also the causes of the majority of deaths in houseplants.) Meadow species differ in their ability to tolerate waterlogging. Drought during the growing season is another problem that some species are better able to tolerate than others. David had found that there was a close correlation between the abundance of different species at different locations in the meadow and the information his map gave him about the hydrology of those locations. He drew graphs of where each species was found in relation to two quantities: how long it spent in waterlogged conditions and how long its growing places were droughted. There was a graph for the distribution of three species of buttercup and another showing how five different species of sedge were distributed in relation to degrees of waterlogging and drought.

Immediately these graphs flashed up on the screen it was as though the darkened seminar room had been flooded with light. These graphs were maps of niche space, and they quite clearly showed that species in the same genus were differently distributed! My excitement was such that I could scarcely sit still through the remainder of the seminar. Had plant niches been found at last?

To understand more clearly how a map of niche space is related to a map of physical space, think about a more familiar kind of map that also presents an abstraction. The London Underground (Tube) map, devised by Harry Beck, is a classic piece of design whose concept has been copied by metros and subways throughout the world. What was different about this map when it first came out in the 1930s was how it translated the physical locations of stations into their functional relationships for a traveler. It shows you which stations are connected to each other in traveler-space, which is what you really need to know to help you get from *A* to *B*. You do not need to know that Knightsbridge and Covent Garden are two miles apart, but what it is useful to know is that you can get from one to the other by traveling four stops on the Piccadilly Line.

David's graphs of plants in niche space operate in a similar kind of way. They show you the distance between species in functional terms, rather than in physical proximity. Just as in the Tube map, physical space and functionality are related to each other. Stations that are physically far apart are functionally connected if they are on the same Tube line. Stations that are physically near each other are functionally separated if no Tube line runs between them. Many English meadows, like Cricklade, have a gently undulating topography with long, broad ridges that alternate with shallow depressions or furrows. In such a meadow adjacent plants can experience quite different ecological conditions if one occupies the top of a ridge and the other the bottom of a furrow. The ridge-top species experience dry conditions most of the year, while the furrow-bottom species are flooded part of the year and wet for much of the rest. The species that compete most strongly with one another are those that prefer the same conditions. Our map of meadow plants shows us what the ecological relationships among species are in niche space.

It has long been known that the three species of buttercup found in meadows tend to occupy different locations in ridge-and-furrow fields, with bulbous buttercup perched on the ridges, tall buttercup preferring ridge sides, and creeping buttercup most abundant in furrow-bottoms. Before David Gowing's study, this textbook example of niche separation in plants appeared to be unique. Everyone thought that buttercups were the exception rather than the rule because other species segregate much less visibly. By using a sophisticated method of measuring variation in soil moisture over a fifteen-year period, David and his colleagues were able to map plant niches in a new way. The result was like turning night into day. Where before we had been blundering around under a dimly lit street lamp, now everything was brilliantly illuminated.

David had arrived at his discovery through applied research and when I heard his talk there were still some fundamental questions that needed answering before we could be sure that his niche maps were really showing us what I believed they were. There seemed little doubt that he had discovered how to map the niches of meadow plants, but what we did not know at that point was whether the eighty species typically found in meadows really occupy significantly different zones in niche space. Perhaps their apparent tendency to avoid each other on the niche map was no more than an illusion, like the signs of the zodiac that the mind of an astrologer imposes on the night sky. How do you tell ecological reality from imaginary patterns like Taurus that are just bull? Luckily, this problem has long been familiar in animal ecology and there are mathematical tools to solve it. The tools had hardly ever been used on plants before because no one until now had found a way of putting numbers to plant niches.

David Gowing, Mike Dodd, and I decided to collaborate on the analysis of the niche structure of meadow plant communities. The first thing we did was to construct a computer null model (chapter 5) from the data by randomizing the distribution of species in niche space over and over again. Each time we made a random meadow, we calculated the average niche overlap, which told us how far apart on the niche map all possible pairs of species were. This was like randomizing a map of London, then measuring how far apart along Tube lines the new Knightsbridge and Covent Garden stations were. The result was amazing. It showed that species in the real meadow were *much more* spread out in niche space than you would expect by chance. It meant that buttercups were not the exception, but the rule. Here was the first strong evidence ever found of niche separation occurring throughout an entire plant community!

A good rule in science, and indeed in life generally, is that if something is too good to be true, it probably isn't. So, we checked and double-checked our data, and then did the same kind of analysis for an even bigger dataset that David had collected from another meadow at Tadham, in Somerset. The meadow at Tadham is as flat as a pancake, without any undulations, but we got the same result. Here, soil moisture varied not with topography as at Cricklade, but with distance from drainage ditches. Now we had found how to measure them, these niches just wouldn't go away! The next question was "Why?"

Niche separation between species means that they are specialized to different environmental conditions or resources. Specialization arises when trade-offs make it impossible to succeed as a Jack-of-all-trades (or a Jill-of-all-environments) when competing with other specialists. This is why different species of Darwin's finch (chapter 5) have different beak sizes: big beaks are needed for large, tough seeds but are inefficient for gathering enough small seeds to support a big-beaked bird. Small-beaked birds cannot tackle big seeds, but can gather enough small, soft seeds to make a living. What is the trade-off in meadow plants that drives specialization? Of course the answer had to be found in the niche maps. The maps were drawn from two measurements: time in waterlogged conditions and time in drought. One might expect that these quantities would simply be the inverse of one another, and that where one is high the other must necessarily be low, but they are defined in such a way as to avoid this. There are large regions of niche space that are neither particularly waterlogged in spring nor particularly droughted in summer.

The obvious trade-off to look for was between tolerance of waterlogging and tolerance of drought, so we worked out average scores of these quantities for each of sixty different meadow species and then simply plotted the scores against each other. The points for the sixty species fell along a diagonal line with a downward (negative) slope. This meant that there was indeed a strong trade-off: species could be well adapted to moist conditions or to dry ones, but not to both. The story was now almost complete, but there was one last thing to investigate before we could publish our results. We wanted to know whether the trade-off was a universal one that applied to all plants, or whether perhaps there was just a contrast between the monocots, which include many waterlogging-tolerant species, and the eudicots that might prefer drier conditions. To find that the trade-off really only had two data points, monocots at the wet end and eudicots at the dry one, would weaken the generality of our results and we had to check for this. We used the tree of trees (chapter 2) to incorporate the phylogenetic relationships among species into our analysis and again tested for the trade-off. It was still there! We now had strong evidence that the trade-off that produced niche segregation in meadow plants was widely distributed among flowering plants. The implication of this is that niche segregation should not be confined to English meadows because the trade-off on which it is based must occur in other species too. We had found an ecological principle that might apply to many different kinds of plant community all over the world.

In a fever of excitement I wrote up our results and mailed the paper off to the scientific journal *Nature*. The prestige of this journal is such that every scientist wants to publish in it and over 90 percent of the manuscripts the journal receives are sent straight back to the authors without even being refereed by other scientists who are experts in the field. I waited a week with baited breath, expecting by every post to receive that familiar letter from an editorial underling at *Nature* containing the killer phrase "of insufficient general interest."

PLATE I.

Succulent plants in the Princess of Wales Conservatory at Kew Gardens (chapter 1). Photograph © Mike Dodd.

PLATE 2.

A wild dragon tree and other endemic plants on the island of Tenerife in the Canary Islands (chapter 3). Photograph © Mike Dodd.

PLATE 3.

Fir waves at Mount Shimagare in Japan (chapter 4).
Photograph © Mike Dodd.

PLATE 4.

A species-rich English meadow (chapter 6).
Photograph © Mike Dodd.

PLATE 5.

The world's oldest ecological experiment, the Park Grass Experiment, at Rothamsted (chapter 7). Rothamsted Manor can be seen in the top right corner. Photograph by Richard F. Wallis Photography, Bedford, UK, used by permission of M. J. Hawkesford, Rothamsted Research, UK.

PLATE 6.

A pine rockland community at the Hole in the Donut in the Everglades (chapter 8). Photograph © Mike Dodd.

PLATE 7.

A field of canola (oilseed rape), one of the crops that has been genetically modified for herbicide tolerance (chapter 9). Photograph © Mike Dodd.

PLATE 8.

An English lowland heath, with gorse (yellow) and heather (purple) (chapter 10). Photograph © Mike Dodd.

It didn't come. We were over the first hurdle, which was to get our paper considered by referees. About six weeks later we received the verdict of the three anonymous referees. One said "This is the plant equivalent of Darwin's finches." A second raised some technical questions, and the third damned us with faint praise. They said it was all very well to show that the species had different niches, but we needed to do an experiment to prove that this was really caused by competition. Quite reasonably, the editor's judgment was that we needed to answer these points before he could accept the paper, but the last referee seemed to have raised an impossible obstacle. We privately dubbed this anonymous referee "Rumplestiltskin."

We had fifteen years data on hydrology, and results that had never been seen before in any plant community. What feasible experiment could possibly prove something that already seemed so obvious and new? As I thought about this problem I realized that we were in luck: a suitable experiment had already been performed over forty years previously. In the 1950s, the German ecologist Heinz Ellenberg had constructed a large concrete tank that he filled with soil and graded into a slope so that the soil was deep at one end of the tank and shallow at the other. The tank was watered from beneath, so that the water table was over a meter beneath the surface at the deep end but quite near the surface at the other. This arrangement created a water-table gradient just like the gradients experienced by plants in meadows. Ellenberg then took the seeds of meadow plants, including six species also found in David's meadow samples, and sowed them in strips along the length of the tank. In one strip Ellenberg sowed a mixture of all the meadow species, and then in separate, parallel strips he sowed each of the species on its own, one species to a strip. The plants grown in the strip containing a mixture of species competed with others that had a variety of tolerances to waterlogging and drought. Those plants growing in strips of a single species competed only with their own kind.

After a year, Ellenberg harvested the plants in sections down the gradient, and weighed the growth made by each species when grown in mixture and when grown alone. He used his results to draw graphs of how well each species grew at different locations along the water-table gradient. In the strips where each species grew on its own, all of the species did best near the middle of the gradient. If you think of the gradient as a line drawn through niche space, then all the species congregated in the same niche when they did not have to compete with each other. By contrast, when growing in the strip where species were mixed up together, each species did best at a *different* location along the gradient. Competition between species forced them into different niches. Ellenberg's experimental results supported our own from the field perfectly and fully justified the comparison with Darwin's finches, which also show niche-shifts where species compete (chapter 5).

Ellenberg's results had long been neglected in the English-speaking world, possibly because they were published in German, but also I suspect because the corresponding field data that were needed to show what happened in a natural community were lacking. Now, field and experimental results could come to each other's aid. We certainly needed the help! I recovered Ellenberg's raw data from his published graphs and re-analyzed them using the same mathematical tools we had used on David's field data. We calculated the degree of niche overlap between species grown on their own and compared this quantity with niche overlaps for ten thousand randomized versions of the data. This analysis showed that niches overlapped so much when the species were not competing with one another that the chance of getting a result like Ellenberg's by chance was less than one in ten thousand. Not one of the randomized datasets showed as much niche overlap between species as the actual overlap observed in Ellenberg's single-species strips. The message was clear: when not competing with one another, species are not scattered randomly in niche space, they all crowd into the same sweet spot in the middle.

This result was very important because it told us that, by comparing our field measurements of niche overlap with random patterns, we had performed a highly conservative test. Ellenberg's experiment showed that the proper point of comparison for field data is not a random distribution of species, but a clumped one. Competition between species in the field might drive species niches far enough apart for them to look random rather than clumped, but our test would not detect that this had happened if we used random distributions as a benchmark. We would wrongly conclude that such distributions did not show niche separation, when in fact it *was* there because plants in the same niche ought to be clumped. The implication was that we had underestimated the degree of niche segregation in meadows. It must be even stronger than we had thought.

When we calculated the niche overlap between species in Ellenberg's mixture strip and compared it to the null model based on the same species growing on their own we found a huge reduction in overlap. Competition between the species on the water-table gradient had such a powerful effect in driving species apart that the overlap between them dropped off the bottom of the scale set by the randomized data. Rumplestiltskin had done us a favor because we had passed the challenge with flying colors. The new results were incorporated into the manuscript and back it went to *Nature.*

Publication in *Nature* of the first clear demonstration of community-wide niche segregation in a plant community was not the end of a story, but only the beginning. There is tantalizing evidence that soil moisture is the key to niche separation in many other plant communities, such as between different oak species in Florida, but it is still too early to tell how widespread the phenomenon really is. David Gowing and his students are now investigating *how* competition causes species to occupy different niches on hydrological gradients. That will be a story for the future.

The dual keys to botanical diversity in most grasslands are grazing by mammals and niche specialization by plants. The first of these conclusions is clear, but we are only just beginning to understand the second. The hydrological niches of meadow plants are just one aspect of plant niches. Other trade-offs, such as between competitive and dispersal abilities (chapter 5), create further kinds of specialization that need to be included in an enlarged niche concept. What is certain is that grasslands, like all plant communities, shelter fettered demons that would smother their neighbors at the first opportunity. Remove herbivores, change the drainage, or add nutrients and demon grasses take control. Of all these demon-inducing threats to plant diversity, it is added nutrient that is by far the most widespread, and it is issuing from a surprising direction, as we discover in the chapter that follows.

7

Liebig's Revenge

A light drizzle of honeydew moistens the air. Aphids in the lime trees overhanging our path puncture the leaves with a billion stylets and the sap surges through each miniature thief, spilling sugared water to the ground like ale from a tiny spigot. Even billions of wastrel insects cannot diminish the grandeur of this avenue of tall, arching boughs that meet to form a glowing green canopy far above our heads. Its breadth and stature suggest that the avenue leads to something grander than an ordinary farmhouse, and sure enough, at the end of the single-track road are crested iron gates, set obligingly open to receive us. Inside is Rothamsted Manor, its façade of old red bricks and mullioned windows radiating warmth in the June sunshine; the tower clock perched on the tiled roof shows that it is nearly noon. This is good haymaking weather, something that cannot always be relied on in England in mid-June.

The drone of a tractor can be heard nearby, a sound you might normally shun, but this particular machine is harvesting something rather special—in fact, something unique. It is the 150th annual harvest of the Park Grass Experiment—an event we will be privileged to witness. We follow the gravel drive that crosses in front of the manor house and head through a small spinney toward the noise of the tractor. In front of us is a large meadow divided across its width into strips, each about twenty meters wide. This is our first glimpse of the world's longest-running ecological experiment.

Let's take a closer look. Here is Plot 3, a now scarce fragment of species-rich hay meadow vegetation. Back in 1856, when the experiment was started, the whole field was like this and every farm in the neighborhood had meadows just like it. The plot is a tapestry of different textures and colors, with the warp of the fabric formed by a stiff, wiry-looking grass—red fescue—that features long, graceful flower stems. Another half-dozen grasses lend subtle variations of color and texture to the background, but the highlights are broad-leaved plants like birds-foot trefoil, ladies' bedstraw, and fairy flax—species also found in chalk grassland (chapter 6). Take a step back from Plot 3 and it is easy to see, even at a casual glance, that other plots in the Park Grass meadow are quite different. A few plots to our left, for example, is an area that receives an annual dose of fertilizer containing nitrogen and other minerals. In contrast to Plot 3, which has had no fertilizer for 150 years, there are very few colorful flowers, the grasses are taller, and the vegetation seems more bulky. At the beginning, in 1856, this plot had as many species growing on it as Plot 3, but the fertilizer applied to it increased the abundance of grasses at the expense of other species. The result was the now familiar story of a few demon species supplanting the rest when added nutrients gave them the opportunity to do so (chapter 5). The Park Grass Experiment is a microcosm of the larger world of plant communities, where we can see, enacted in fewer than three subdivided hectares, how soil nutrients hold the ring in the battle between plants, the outcome of which determines species diversity. The experiment is also the living trophy of another kind of battle, one among three giants of agricultural science in the nineteenth century.

The fertilizer treatments applied to the Park Grass meadow were begun by John Bennet Lawes and his assistant Joseph Henry Gilbert as part of the larger scientific research program that they devised to investigate the effects of chemical fertilizers on the yield of different crops. In the days when horses drew the plow, pulled the first agricultural machinery, and carted the harvest, hay was integral to the agricultural economy of every farm. Farmers were just as interested in a heavy crop of hay to feed their horses as in a good crop of wheat or barley, and Lawes was in business to help them achieve this. An advertisement in the *Gardener's Chronicle* for July 1, 1843 stated:

> J. B. LAWES'S PATENT MANURES, composed of Super Phosphate of Lime, Phosphate of Ammonia, Silicate of Potass, &c., are now for sale at his Factory, Deptford-Creek, London, price 4s. 6d. per bushel.

Lawes's matter-of-fact advertisement has, to modern sensibilities, the economy of a small ad in the personal column rather than the attention-grabbing force of the glossy, full-page ads that sell fertilizers today. Nevertheless, Lawes's superphosphate business was highly profitable, providing him with the funds to build the world's first scientific agricultural research station on his family estate at Rothamsted in Harpenden, a half-hour train journey from central London.

Seven field-scale experiments begun by Lawes and Gilbert are still running at Rothamsted, of which the Park Grass Experiment was the last to be initiated. So, by 1856 when the Park Grass meadow was divided into plots and chemical fertilizers were first applied, Lawes and Gilbert had a pretty good idea of what would happen to hay yields. But they very soon noticed something else that they had not expected: "[T]he plots had each so distinctive a character in regard to the prevalence of different plants that the experimental ground looked almost as much as if it were devoted to trials with different seeds as with different manures."

Botanical sampling to record the differences between treatments was designed and executed with a rigor and meticulousness that came to typify scientific research at Rothamsted. As the mower scythed his way across each plot, he was followed by a trained botanist who gathered representative handfuls of hay, which were passed behind to a boy carrying a sack. The hay samples were then dried and stored for later scrutiny in the laboratory. Under close supervision from a botanist in the lab, boys sorted through each sample of dried hay, teasing it apart and placing each leaf into its correct pile, one for each species. The unidentified fragments and seeds that were left at the end of this process were further sorted and then each completed pile was weighed to determine how much of each species there was in the hay from each plot. This task took ten months to finish and, as might be imagined, severely tested the patience of the boys undertaking the work. One of the most diligent of them, a lad named Edwin Grey, was first employed at the age of twelve to separate the grass samples collected in 1872. He went on to become a permanent employee and worked at Rothamsted for fifty years, eventually rising to field superintendent. In his memoirs, he recalled the 1872 botanical separation:

> Some plots were much more difficult to botanise than others. One boy named Swallow, whose place was next to me at the table, somehow whilst working at a particularly difficult plot, always managed to get his small and broken pieces finished first. He was praised for being so quick with his work, but we other boys did not believe that he could get through so much quicker than us by fair means, so we watched, and after a time discovered that he threw a considerable part of his small pieces into the earth closet at the back of the building.

Swallow was instantly dismissed. The observations of botanical composition at Park Grass and the annual treatments continued, and the record now entitles this meadow of seminatural vegetation to claim to be the world's longest-running ecological experiment. Why, though, did Lawes and Gilbert oblige posterity by keeping Park Grass and the other experiments running beyond the first few years? They had plainly established that chemical fertilizers were as good as farmyard manure in improving yields within the first decade of all of the experiments. So why continue to expend the enormous effort and expense involved in maintaining and monitoring them? This question has intrigued many people, not least the scientists who work at Rothamsted today. A. E. "Johnny" Johnston, formerly of the soil science department at Rothamsted, has speculated that a long-running scientific argument between Lawes and Gilbert and the German agricultural chemist Baron Justus von Liebig may be a part of the explanation.

Liebig was the foremost authority on agricultural chemistry of his day. He was a towering intellectual figure who trained a whole generation of chemists in his laboratories, including Joseph Henry Gilbert, who studied with him in Giessen, Germany. Liebig's preeminence in agricultural chemistry was recognized by the British Association for the Advancement of Science, which invited him to prepare a report on agricultural chemistry, the first part of which was published in 1840. Such was Liebig's celebrity that when he visited Britain in 1842 and presented his passport for inspection in London, the passport official shook him by the hand and exclaimed how pleased he was to meet the author of *Agricultural Chemistry*, which he had read. On a later visit he was the personal guest of Queen Victoria and Prince Albert at Balmoral Castle in Scotland.

Liebig wrote that proper husbandry had to be based on an understanding of chemistry. In order to grow, plants require a variety of elements such as nitrogen, potassium, phosphorus, and magnesium. The minerals that are required for plant growth, Liebig reasoned, are those revealed in the composition of the ash that remains after plant material is burnt. Nitrogen escapes when plants are burned and is not found in the ash, which led Liebig to underestimate its importance to plant growth. Land is depleted of these vital minerals when crops are harvested and unless they are returned in some form to the soil, it will become impoverished: "It must be admitted as a principle of agriculture that those substances which have been removed from a soil must be completely restored to it, and whether this restoration be effected by means of excrements, ashes or bones, is in great measure a matter of indifference."

What did matter was the amounts of each mineral element available in the soil, and especially the amount of the least available element because this would be the one that limited plant growth. Though Liebig was not the first to expound this idea, it became known as, and is still called, "Liebig's Law of the Minimum." The law is fundamental to plant growth, but it is as simple as the notion that a chain is only as strong as its weakest link. If each link in the chain represents a different element, and all are required for growth, then the element that is exhausted first will be the one that causes the chain to snap and stops plants from growing. Of course, applying a fertilizer that contains the element in shortest supply will promote growth and some other element may then become limiting.

One of the questions that Liebig addressed in his book concerned where plants obtain nitrogen, and here he made an uncharacteristic error—one that would take him more than twenty years to acknowledge. Nitrogen is a chemical element essential to life and is required in large quantities for the manufacture of proteins and other molecules by all living things. Animals obtain their nitrogen directly (in the case of herbivores) or indirectly (in the case of carnivores) from that contained in plant tissues. Therefore, understanding how plants get their nitrogen is essential to understanding the agricultural production of livestock as well as crops. Eighty percent of the air we breath is nitrogen gas, but the element is chemically inert (and odorless) in this form and cannot be assimilated, so it all goes out again with every exhaled breath. Plants cannot assimilate nitrogen gas either, but they can take up various compounds of nitrogen by way of their roots—in particular nitrate, which is highly soluble, and ammonium, which is produced when ammonia gas dissolves in water. (Ammonia is the gas that gives stale urine its pungent smell.) Liebig stated in *Agricultural Chemistry:* "No conclusion can then have a better foundation than this, that it is the ammonia in the atmosphere which furnishes nitrogen to plants."

Even in 1840, this statement was at odds with the experience of farmers who knew that adding nitrogen-containing manures to soil improved crop growth. In terms of Liebig's Law of the Minimum, nitrogen is usually the element in shortest supply, and therefore applying nitrogenous fertilizers has a large impact on plant growth. The reason that nitrogen, of all the elements, is so often limiting to growth is simply that plants require a great deal of it. Three years later, in a new edition of his book that Liebig published in 1843, he strengthened his views by adding that whatever the cause of the beneficial effects of manure on crops, it "cannot be due to its nitrogen."

John Lawes and Joseph Gilbert began their fertilizer experiments in 1843; four years later, Lawes would contest Liebig's conclusions in a paper he published. Liebig's riposte came in 1851 when he dismissed Lawes's experiments as "entirely devoid of value as the foundation for general conclusions." In private, he later wrote to an English supporter in even stronger terms: "[I]t is all humbug, most impudent humbug. . . . Lawes and Gilbert hitch on to me like a vile vermin and I must get rid of them by all means."

Further exchanges followed, with Liebig eventually claiming that he had been misunderstood and Lawes asserting that Liebig was equivocating in the face of the mounting experimental evidence that plants obtain their nitrogen from the soil rather than from the atmosphere. If one appreciates the obduracy as well as the enormous influence of Baron Justus von Liebig, it is easier to understand why Lawes and Gilbert might have been keen to maintain their fertilizer experiments as living proof of their side of the argument.

Lawes and Gilbert were right about the source and importance of plant nitrogen, but that is not the end of the story and the baron had his revenge long after the protagonists in the argument were all dead. Edwin Grey records in his memoirs that early one morning in September 1916, during World War I, he was lying in bed in Laboratory Cottage at Rothamsted when he heard a German Zeppelin airship go overhead and drop a bomb on Lawes and Gilbert's wheat experiment in Broadbalk field. This made a crater in Plot 6, the treatment receiving a small amount of all the main minerals. "Now why on Earth did the Germans bomb Broadbalk?" Grey wondered. Was this Liebig's revenge? No, of course not, but Liebig's revenge did come from the air, and, as we shall presently see, it was more subtle, more pervasive, and much more devastating to plant diversity than a World-War-I-vintage bomb.

The noise of the tractor working its way across the plots is getting louder and brings us back to the present, reminding us that we need to catch up with more recent research at Park Grass. Modern ecologists have learned a great deal from the Park Grass Experiment and it has inspired scientists throughout the world to emulate it, notably David Tilman of the University of Minnesota, who in the 1980s constructed vast experimental arrays of grassland plots at Cedar Creek in Minnesota. He has said of Park Grass that "[t]here are no other long-term studies of this kind in existence . . . it's like a well that ecologists can dip into whenever they want to test a new idea."

I first dipped my own small bucket into the well in 1977 when I was a graduate student, looking for a way to extend the limited horizons of three years of doctoral research to something more realistically like the long time spans over which plant populations change. When I heard about Park Grass, I couldn't contain my excitement and immediately phoned the library at Rothamsted to obtain all the annual reports in which, year by year, the results had been published. For a bargain price, affordable even on a tight budget, I obtained all the data including a broken-backed, leather-bound copy of Lawes and Gilbert's first major scientific papers on Park Grass published a century before. I analyzed the relationship between the average hay yield of each plot and the diversity of species found in its hay and discovered that the two were remarkably closely correlated. Looking at the data from the first full botanical separation that was made in 1865, I found that each extra metric ton of hay per plot yielded appeared to reduce its diversity by at least three species. Unfertilized Plot 3 had a low yield and nearly fifty species on it, while the plot with the highest yield had only twenty. Data from later harvests showed the same pattern. There are few solid laws in ecology, but this seemed to be one of them. You could accurately predict the species diversity of these grasslands from their average hay yields. There was a complication, however, and a very interesting one at that.

One of the treatments applied to certain plots in the Park Grass Experiment is a nitrogenous fertilizer called ammonium sulphate. This supplies nitrogen that boosts hay yield, but it also greatly acidifies the soil. The result is that the soil becomes progressivley more acid with each application of the fertilizer. Before the tractor gets to it, let's take a quick look at Plot 9, which has received ammonium sulphate annually for the last 150 years. It will immediately strike you that there is only one species on this plot. It's a rather brittle-looking plant called sweet vernal grass (*Anthoxanthum odoratum*). Crush a leaf and it releases the odor of coumarin. It is the smell of this species that gives new-mown hay its characteristic fragrance. This plant occurs in varying amounts on virtually all plots in the Park Grass Experiment. Looking back over the history of the experiment, as the soil on certain plots became more and more acid with time, one species after another disappeared, till only *Anthoxanthum odoratum* was left on some of them. In conditions of extreme acidity, the textures, the attractive flowers, the diversity, even the earthworms and other soil inhabitants are all entirely gone. All that remains is the lingering odor of coumarin, the essence of hay meadow having become a sweet vernal ghost.

My analysis of all the Park Grass data chronicled the effect of progressive acidification and showed that this had added to the detrimental effects of fertilizers on species diversity. In any one year you could see the same relationship between hay yield and diversity in acidified and in unacidified plots, the only difference being that every acid plot paid a fixed extra penalty in lost species, and the size of that penalty became larger as time went on. In 1862, before acidification really began to bite, the acid penalty was undetectable, but fifty years later every acid plot had scarcely half the species of a non-acid plot with the same hay yield.

These results showed that the fertilizer treatments at Park Grass had two separate kinds of negative effect on diversity. First, any fertilizer treatment that increased the total yield of hay would quite quickly lead to an exactly corresponding reduction in the number of species present in the plant community. This is because adding nutrients favors the few species best able to exploit them, and the growth made by these demons increases total yield while outcompeting other species. If the fertilizer contains nitrogen, a few competitive grasses are the species favored, but if nitrogen is absent and only phosphorus and potassium are added, plants of the pea family take over. Either way, it is demons versus diversity.

The second mechanism operates in an entirely different way and has less to do with competition than with soil chemistry. Simply put, most plants accustomed to living in neutral soils have a hard time dealing with acidity, and fewer and fewer species are able to cope as acidity increases. Interestingly enough, *Anthoxanthum odoratum* can probably only cope at Park Grass because it evolved tolerance of such conditions as soils became more acid.

Fertilizers are not the only source of nitrogen inputs to the experiments at Rothamsted—there are inputs from the atmosphere too. It is estimated that, back in 1850, each hectare of ground received about ten kilograms of nitrogen per year, mostly from compounds dissolved in rain. This was a negligible amount, quite unable to compensate for the nitrogen removed in plant tissues with each annual harvest and having little effect on soil acidity. In consequence, it was probably without any ecological effects on plant diversity. Over time, however, nitrogen inputs have risen until today, about forty-five kilograms per hectare is deposited annually—a very significant amount. Most, if not all, of the increase is from the burning of fossil fuels, which produce various airborne oxides of nitrogen. Car exhaust gases are one source and fuel burnt to heat homes another. In summer, when people in Harpenden and other nearby towns turn off their gas-fired central heating, the nitrogen dioxide readings at Rothamsted halve, climbing steadily again as winter approaches. Then, at Christmas and New Year, there is a brief but detectable downward blip in the data as people leave their cars in the garage and give the atmosphere a festive break—a salutary reminder that our everyday activities have a measurable impact on the environment.

Atmospheric nitrogen inputs to ecosystems are now a global phenomenon, found everywhere that fossil fuels are burned or intensive agriculture is practiced. This is Liebig's revenge. Plants now do get significant amounts of usable nitrogen from the atmosphere, though not through any natural process, and these inputs are threatening the diversity of plant communities. The threat is so insidious and all-pervasive that ecologists are only just beginning to realize quite how serious it is. As recently as 2004, Carly Stevens, a graduate student at the Open University, made news headlines when she published in *Science* magazine her discovery that plant diversity in British grasslands has been significantly damaged by atmospheric nitrogen deposition. *Science* is so prestigious a publication that most scientists would count themselves lucky to publish in it once in their entire career, so for a graduate student to pull this off before she had even submitted her thesis was a pretty big deal.

Carly investigated the species diversity of naturally acidic grasslands at sixty-eight sites stretching the length of Britain, from the Highlands of Scotland in the north to the southwest of England. At each site she counted the number of different plant species in each of five large quadrats and sampled soils for analysis back in the lab. She then analyzed the relationship between species diversity at each site and twenty different environmental variables that are known to affect this. It turned out that total nitrogen deposition was easily the strongest predictor of the average number of species present in samples at Carly's sites. In western Scotland, where the air is unpolluted, there were more than twenty-seven species per quadrat, but in the south of England, where the atmospheric deposition of nitrogen is more than thirty kilograms per hectare, counts went as low as seven species per quadrat. As with Park Grass, soil acidity also influenced species diversity, though the effect was weaker than that of nitrogen.

The significance of Carly's study was that no one had realized just how sensitive acid grasslands are to atmospheric nitrogen deposition, or how much damage to the flora of these habitats has already been done by nitrogen pollution. Carly calculated that for every two-and-a-half kilograms of nitrogen deposited per hectare, one more species went missing from her quadrats. Grasslands of this kind occur throughout Europe, Australia, and North America, so it is no surprise that Carly's research made news throughout the world. For a week she walked on air.

There is a double irony in Liebig's revenge. First is the fact that plants are now receiving appreciable amounts of nitrogen from the atmosphere. But, second, is the unfortunate consequence that, because of Liebig's Law of the Minimum, the ecological effects of the extra nitrogen on natural habitats are large. Plant growth in most temperate zone habitats is limited by nitrogen, and communities growing on soils that have especially low levels are particularly sensitive to disruption by its addition. The spread of tor grass in chalk grassland mentioned in the previous chapter is aided by atmospheric nitrogen deposition as well as by its being unpalatable to sheep.

In European lowland heathlands, heather, grasses, and bracken compete with one another, but the sandy soils that support heathland vegetation are usually too nutrient-poor to allow one species to acquire permanent dominance over the others. In Dutch heathlands, however, atmospheric nitrogen pollution in the form of ammonia released by intensive animal husbandry has turned heathlands into grassland. This has occurred by way of an interaction between nitrogen pollution and a beetle that eats heather. When the beetle is present in only low numbers, as it is in most years, additional nitrogen increases the growth of both grass and heather, and heather is able to hold its own because the grasses cannot penetrate the canopy of the heather bushes. But heather beetles, like many herbivorous insects, vary a great deal in abundance from year to year. When they are abundant, the larvae of the heather beetle grow particularly well on a diet of heather shoots that have an enriched nitrogen content. The voraciously feeding beetle larvae and adults defoliate and even kill bushes, tipping the competitive balance between heather and grasses irrevocably toward grasses.

Forest trees are also susceptible to nitrogen pollution. At Harvard Forest in Massachusetts, the effects of atmospheric nitrogen deposition have been simulated in experiments on forest plots of red pine and broad-leaved trees such as oak, American beech, and maple. The initial effects of the added nitrogen were what you might expect in a nitrogen-limited forest: the trees grew better. But after fifteen years of exposure to additional nitrogen inputs similar to those found in polluted areas, the forest ecosystem had become saturated with nitrogen, changing the chemistry of the soils. The canopies of the red pines had thinned and lost foliage and a great many trees had died. In the polluted broad-leaved plots, red maple was also especially vulnerable. These effects are not just occurring in experiments. In the forests on Whiteface Mountain in the Adirondacks, where we visited in chapter 4, as well as on other mountains in the region, red spruce is disappearing as it succumbs to atmospheric nitrogen pollution to which it seems to be particularly susceptible. Balsam fir is taking its place and increasing its dominance in these forests.

Atmospheric nitrogen deposition has historically been much worse in Europe than in North America, but American ecologists are beginning to discover the possible consequences of a worsening situation in a variety of their terrestrial ecosystems. The Mojave Desert in the American southwest is a nitrogen-poor ecosystem that is becoming exposed to atmospheric nitrogen deposition from the spread of nearby urban areas. Experimental nitrogen additions to desert plots suggest that nitrogen pollution will enable nonnative plants to invade the desert, exposing the diverse desert flora of the Mojave to the threat of competition from plants that they may be unable to resist. Plants of nitrogen-limited habitats such as the Mojave are generally adapted to those conditions and cannot compete with vigorously growing invaders from more nitrogen-rich environments. All that keeps the invaders out at the moment is the current paucity of nitrogen in the soil.

Nitrogen pollution can change an ecosystem from a pristine state where nitrogen limits growth to one where phosphorus or some other nutrient is limiting. This change can favor invaders and may be what has happened to the native grasslands of the western United States and Canada where, among other alien plants, no fewer than five nonnative species of knapweed belonging to the genus *Centaurea* dominate pastures. These spiny and unpalatable weeds are sometimes so dense that they make grasslands totally unusable by livestock. Before invasion, the growth of western grasslands was nitrogen-limited, but it is thought that ammonia pollution from feedlots, where cattle are kept on an industrial scale, changed this. In an experiment, the addition of nitrogen fertilizer to a stand of diffuse knapweed (*Centaurea diffusa*) failed to increase the size of plants, but plants fertilized with phosphorus were more than twice the size of unfertilized plants. This indicates that phosphorus rather than nitrogen now limits plant growth. *Centaurea* species seem to be better adapted to this altered situation than native grasses are. The consequence is a loss of native plant diversity as well as of grazing land.

Liebig's revenge—the artificial fertilization of natural habitats all over the industrialized world with atmospheric nitrogen pollution—is like a massive, unplanned, and reckless experiment in plant nutrition. It has revealed that many natural plant communities are nitrogen-limited. In habitats like British grasslands, release from nitrogen limitation has unleashed the demonic potential of native grasses, and competition from these has eroded plant diversity. Elsewhere, particularly in the New World, nitrogen pollution has opened natural communities to invasion by nonnative plants. Many plants have a habit of running amok when they are introduced to new environments. This is when some plants really earn the title of "demon," invading and changing whole plant communities, as can be seen at our next stop in Florida.

8

Florida!

The very name of the state of Florida evokes its botanical riches. Plants, like humans, revel in its subtropical climate. Florida's forests, swamps, and freshwaters harbor more than four thousand plant species, but nearly a third of them were introduced, not native. In 1920 a prophetic naturalist, Charles Torrey Simpson, foresaw the problems of introduced plants: "[T]here are the adventive plants, the wanderers, of which we have, as yet, comparatively few species; but later, when the country is older and more generally cultivated, there will surely be an army of them." The army has arrived with a vengeance and is advancing mercilessly across Florida's natural ecosystems, taking no prisoners. In the United States, only Hawaii has a worse problem with invasives than Florida. The problem is caused by only a small minority of the twelve hundred nonindigenous plant species, but this minority contains some real demons.

The army of alien plants is made up mainly of conscripts, not volunteers. At least 90 percent of the nonnative species were deliberately introduced into the state as ornamental plants, as new crops, or for other purposes. The kudzu vine is an example that is well known because it is a problem throughout much of the South, where it is estimated to cover 2 million acres of forest land alone. The epithets by which it is commonly known in the region tell the story: "the vine that ate the South," "mile-a-minute-vine," and "foot-a-night vine." At the height of the growing season kudzu *can* actually grow a foot in twelve hours and southerner's joke that you must close your windows at night to stop the vine getting in. Abandoned rural buildings can quickly disappear under a blanket of kudzu, but much worse, so do whole forest stands. The U.S. Department of Agriculture declared the plant a weed in 1972, though it still has its defenders who use it for everything from brewing tea to making baskets. For these stalwarts, a real virtue of the plant is that supplies are free and inexhaustible. Kudzu evokes both loathing and affection in the South, though hardly in equal measure. The ecological damage caused by the other sixty-one plants on a list of the most unwanted compiled by the Florida Exotic Pest Plant Council can induce only loathing.

Let's begin our trip to Florida in pristine habitat at Archbold Biological Station, near the heart of the Florida peninsula, at a point roughly equidistant from the cities of Miami, Tampa, and Orlando, and about as far away as it is possible to get from the sprawling growth of urban Florida. It is a special place where some natural habitat still survives amidst the orange groves and ranchland, and where Archbold commands a two-thousand-hectare tract of land dedicated to research and conservation. We accompany Eric Menges, the station's resident plant ecologist, a few steps from his laboratory, across a forlorn railroad track, and into the plant community that he has studied for the last fifteen years—Florida scrub, a sandy plain filled with shrubs, knee-high palmettos, and scattered slash pines. The soil beneath our feet is a pure-white sand that gleams in the early morning light.

"This area burned last year," Eric says, "and it's looking real good. Most plants survived the burn." Fire is a natural element in this ecosystem and the plants are adapted to it. It's amazing how green everything looks, the fresh green leaves of the palmettos and the new leafy branches of the shrubs hiding the charred sticks that remain from last year's burn. Notice your clothes and shoes; a prankster seems to have taken a charcoal pencil to them, sketching random strokes from thigh to sole while your attention was diverted by the spectacle of this phoenix plant community.

"It takes only two years to get full leaf cover back," Eric continues, much to my amazement. I stoop to grab a handful of sand and let it fall through my fingers. It is flecked with charcoal, but otherwise it looks like pure silica. Where are the plants getting their nutrients?

"These are dunes, about a million years old," Eric says, as he kneels and gently prizes a hole in the sand surface. He points to a grey layer that lies about five millimeters beneath the top of the hole. "There's a whole community of organisms in this crust. It includes cyanobacteria that fix nitrogen from the air. We've used tracers and found that the nitrogen fixed in the crust gets into the vascular plants."

It seems that the sand transmits enough light, not only for the cyanobacteria, but also for small plants: "We've found that rosemary seedlings germinate below ground and spend their first year under there too." The Florida rosemary is an evergreen shrub with tiny, fragrant leaves that is characteristic of Florida scrubs. As we walk back to the station we see a rare sandhill crane probing the mud of a temporary marsh for food. Pausing beneath some slash pines, I ask Eric whether the scrub is threatened by any of Florida's army of alien plants. Perhaps detecting a note of disappointment in my voice at not seeing any of the notorious demons, he is almost apologetic in his reply.

"No, though there are some alien species around the station buildings. They can't survive out here, its too tough for them. Not enough nutrients, and they aren't adapted to the fire regime. Cogon grass can be a problem, but we've got that under control here."

At that moment, two scrub jays arrive, chattering in the branches over our heads. They pose and dip in our direction, evidently interested in our presence and quite unafraid. The birds sport leg bands and belong to a celebrated population that has a scientific monograph devoted to them, published by Princeton University Press. This Ivy League avifauna expects attention, and gets plenty. As we retrace our steps toward the field station, I notice some familiar plants growing by the railroad track: *Lantana*, a notorious invader from the West Indies that now occurs in every tropical region, and a briar rose from Europe. These plants lurking in the wings are a warning of the fate awaiting the last remnants of Florida scrub, if prescribed burning is not maintained or human activities enrich the soil. Not only the Florida scrub plants, but sand cranes, scrub jays, and the gopher tortoise (with the numerous other creatures that live in its sand burrows) would be lost. Fortunately, at Archbold, the future of the scrub looks secure. The picture is not so favorable for natural habitats farther south.

From Archbold, we take Route 27 and head south for the Everglades. We quickly leave behind the ridge of old Pliocene dunes and descend into geologically newer terrain. The "river of grass," as Marjory Stoneman Douglas evocatively described the Everglades in her poetic book of 1947, once stretched a hundred miles from Lake Okeechobee to Florida Bay. Water from the lake spilled southward toward the sea, nourishing a marshland prairie dominated by sawgrass.

"For sixty miles or so south of Lake Okeechobee the river of saw grass sweeps wider than the horizon, nothing but saw grass utterly level to the eye, a vast unbroken monotony. The grass crowds all across the visible width and rondure of the earth, like close-fitting fur."

Sawgrass was a demon in its natural habitat, but now those sixty miles south of Lake Okeechobee are cultivated and settled, and the lake itself is little more than a reservoir supplying agriculture and the burgeoning population of southern Florida. The remains of the Everglades must make do with what water is left after Floridians have irrigated their fields, filled their pools, and flushed their toilets.

The water from Lake Okeechobee no longer finds its own course to the sea, but is canalized and controlled every step of the way. It is in these canals that we find the first signs of trouble with demon plants. The waterways are plagued by water hyacinth, a native of the Amazon; water lettuce that reached Florida as early as the eighteenth century, and hydrilla, a relatively recent arrival. All three species have rapid rates of growth and overwhelm native aquatic plants, entirely replacing them and reducing oxygen levels in the water with dire effects for fish and wildlife. Millions of dollars are spent annually on controlling them with herbicides and other measures.

Water hyacinth is truly one of the world's worst weeds, able to double in population size in only fourteen days and infesting sixteen states of the United States and fifty-six countries in the tropics and subtropics. The plant is free-floating and has air-filled bladders that make it extremely buoyant. New plants are produced by budding and by seed. This combination of traits allows the plant to multiply and spread with ease, so that it can blanket the surface of an entire lake in one season of growth. Water hyacinth is successfully managed with herbicides but hydrilla, which grows submerged, has not been brought under control.

It is a curious thing that all hydrilla plants in Florida are female and so, for want of mates, produce no seed. However, the plant has two other means of vegetative propagation that serve it very well. Established plants produce small tubers on the roots that can reach densities of six thousand per square meter. As if this were not enough, the plant also produces small fleshy buds called "turions," which drop off the plant and sprout to form new plants. Turions can reach densities of three thousand per square meter. Tubers, turions, rapid growth, and no natural enemies make hydrilla a demon among demons.

Natural enemies can be used to control invasive plants if suitable herbivores or diseases can be found in a plant's native range. Alligator weed from South America is another alien water plant that was once a problem in Florida, but it has been successfully controlled by the introduction of three insect species from indigenous populations in Argentina. This kind of biological control is the ideal way to manage invading organisms, but natural enemies must be tested before release to make sure that they do not attack nontarget species. There are cases where this precaution has been ignored and the introduced predator has attacked native species in preference to the alien target, driving the natives extinct. Where biological control succeeds, it restores the kind of balance between a plant and its natural enemies that is normal in indigenous species. Alligator weed now survives at about 1 percent of its abundance before its natural enemies caught up with it. This quantity is sufficient to maintain a stock of natural enemies ready to seek out and destroy any new infestations of the plant that may appear.

Back on the road, we head further south along Route 27 and in the town of Clewiston, on the southern shore of Lake Okeechobee, we have our first encounter with *Casuarina*, or Australian pine. These trees have been planted as wind breaks, but they spread and grow rapidly to giant proportions, shading out all other plants with their melancholy, evergreen foliage. Once past the town, the casuarinas disappear, but as soon we cross the county line into Broward, at the northern edge of the Everglades, another sinister Australian appears. Paperbark is only sporadic at first, appearing in stretches along the roadside, but as we turn off Route 27 onto the minor road to the Everglades National Park the road is suddenly engulfed on both sides by tall, crowded stands of the tree. I have to stop myself from braking hard, the impression is so sudden, intense, and oppressive.

Tim Low, an Australian ecologist writes in his book *Feral Future* of his reaction to seeing paperbark growing in Florida:

> I drove past dark forests of paperbarks much vaster than any growing in Australia, lining highways for tens of miles. American paperbarks grow straighter and more crowded than ours, with up to 37,000 trees per hectare, forming forests so gloomy that nothing grows inside. It was

> amazing to see a familiar native tree playing the role of supreme villain. These were the worst weed invasions I had ever seen.

The route we are on could well be the same one Tim Low describes—the impression is the same for mile after mile.

Paperbark was deliberately introduced into Florida and broadcast-sown from aircraft over the Everglades in the mid-1930s in a misguided attempt to drain them. Natural areas were bombed with these demons that suck five times more water from the ground than native sawgrass. Back home in Australia paperbark is attacked by a whole fauna of insect enemies, but in Florida nothing will eat it and, perhaps for this reason, trees grow bigger and better than in their native range.

Growing up to two meters a year, trees can reach thirty-three meters and a large one can produce 20 million wind-borne seeds in a year. Trees start to produce seeds when only two years of age: precocity is another demon habit. The spread of paperbark has been accelerated by large-scale alterations to the hydrology of the Everglades, which have increased the frequency of fire. Paperbark is adapted to fire in its native Australia, so burns only encourage it. In these conditions the number of paperbark stems can increase tenfold each year. At the height of the problem, paperbark occupied nearly two hundred thousand hectares (half a million acres) in southern Florida, though some areas were cleared of seed-bearing individuals in the 1990s.

Eventually the paperbarks give way to cultivated fields and as we near the town of Homestead dozens of nurseries raising and selling garden plants line the roadside. Horticulture is responsible for introducing more than its fair share of Florida's catalogue of woe, including a contender for the title of "Worst Demon": Brazilian pepper. This evergreen shrub with bright-red berries forms huge monotonous stands that exclude all native plants. In the mid-1990s it occupied seven hundred thousand acres of central and southern Florida and was still spreading. Brazilian pepper has destroyed the habitats of rare and endangered plants such as beach jacquemontia and beach star, and threatens the nesting habitat of the gopher tortoise in the Everglades National Park. In addition to the usual demon tricks of copious seed production, good seed dispersal, and rapid growth, Brazilian pepper appears to be toxic to other plants.

We have arranged to rendezvous with three graduate students from Florida International University who will meet us at the entrance to the Everglades National Park and show us around. John Geiger, Hong Liu, and Jed Redwine are waiting for us when we arrive. John is crazy about wild plants and has lived in Florida most of his life: "Brazilian pepper is a great climbing tree, with all those branches. I used to love climbing in them as a kid." Hong is from Hainan Island, China and wants to study invasive plant problems when she has finished her doctoral degree. She jokes about how delicious Chinese water spinach is: "I want to introduce it here. It's so tasty!" Unfortunately, it has already has already been introduced in several parts of the state and the plant has proved difficult to eradicate.

Jed has conservation in his blood. His grandfather was a soil scientist in Oklahoma during the dust-bowl years of the 1930s. As a child, Jed heard how his grandparents couldn't eat for seven years without first blocking the gaps around the doors and windows of their house with wet towels to keep out the choking, powder-fine dust that billowed off the ruined farmland of the prairie: "I expect my lifetime to be a critical time. I expect to witness mass extinction. Most all my fellow grad students feel the same way. For me, my research is a continuation of what my grandfather did."

Most visitors to this part of the Everglades either make straight for the boardwalk just inside the entrance at Royal Palm, where there are spectacular views of the wildlife at close quarters, or they tear down the highway to the tip of the peninsula to launch their boats at Flamingo. We will do neither but are going instead to the "Hole in the Donut," a former agricultural area in the middle of the Everglades where there is a really serious Brazilian pepper problem. After a short drive, we pull up by the side of the road and get out. The day is beginning to get decently hot, but it is the dry season and there are mercifully few mosquitoes about. Looking down the road straight ahead, its two sides present an astonishing contrast.

On the left is a solid, impenetrable thicket of Brazilian pepper about five meters high, stretching away into the distance. Living stems of the bushes are interwoven with fallen, dead branches, forming a barrier that an English hedgelayer would be proud to have created, only this barrier is not the depth of a hedge—it is the depth of an entire field. At first, it looks as if there are no other trees in this horrendous thicket, but then John notices something. "Look! Its *Ardisia*. And here, that's guava! Guava forms thickets just like this in Hawaii." Guava and shoebutton ardisia are both shade-tolerant alien shrubs with fruit that many birds just love. Both are sprouting beneath the Brazilian pepper: round one in a three-cornered battle among demons.

On the other side of the road, not ten meters away, is an entirely different scene—a subtropical pine forest. The pines are widely spaced, tall, and graceful with no lower branches, so that their lofty canopies cast only a dappled shade on the plants below. They are southern Florida slash pines, belonging to the same species that we saw in the scrubs at Archbold, only here they are growing not in pure sand, but on solid rock. The two familiar palmettos from the scrubs are here too, but there the similarities end. Instead of the acid sand of the scrubs, the pine rockland, as this plant community is known, grows on solid limestone. There appears to be no recognizable soil, just plants springing direct from the rock, as if secured by bolts rather than roots. I turn over a broken piece of rock cautiously, in case it conceals a scorpion or a rattlesnake, and find a tiny seedling sprouting in a crevice. Its roots are exposed and you can see how they thread their way through small holes in the rock that are like the pores of a sponge. Porous limestone can hold a lot of water, and this must be how the plants here survive. It is natural hydroponics.

John and Hong scour the ground for some of the wildflowers that occur here in abundance. "The flora here is very diverse," says John. "You can get ten different herb species in just a ten-by-ten-centimeter square, and about 15 percent of the flora is endemic." John points to a delicate mauve flower of the rockland ruellia and Hong finds a specimen of the ground cherry with nodding yellow flowers, each with a purple center. These will produce small fruits like a cherry tomato wrapped in a tiny, papery lantern. Jed is searching in vain for a small cycad called the "Florida arrowroot." The cycads are an ancient group of plants related to the conifers. They have very tough leaves and no doubt were once the food of herbivorous dinosaurs. Now Florida arrowroot is rare because so many of the plants have been dug up for their tubers, which are processed to produce a very fine starch that is sold as culinary arrowroot.

The diversity and density of species in the ground layer of the pine rockland is reminiscent of that found in chalk grassland (chapter 6), and the similarity is no coincidence. Both communities occur on thin, calcareous soils whose nutrient poverty limits the spread of demon plants that would otherwise crowd out the many small, slow-growing species.

But what about the demons next door? Why are they growing there? John explains that more than half a century ago, the native plants were torn down and a rock-crusher was used to chew up the surface, producing a "soil" of limestone gravel. Fertilizers were added and corn was grown in the middle of the Everglades. Now the fields have been abandoned, but the damage is done and demons have taken control.

There must be a huge rain of Brazilian pepper seeds into the remaining natural habitat from just across the road. Why haven't they invaded it? "We do sometimes find seedlings," Jed says, "but low nutrients and regular fire keep them out. This particular site is burned every two years." Once again, the importance of wildfire.

Florida's natural habitats are supremely elemental—they are bathed in water, nourished by rock, and protected by fire. Where fires have been prevented in the Everglades, or where nutrients have been added such as in the former fields in the Hole in the Donut, Brazilian pepper invades. Demons often need feeding, and this particular one is not well adapted to fire.

Understanding how Florida's natural ecosystems work is essential to protecting them from invasion. Unfortunately, there are alien plants that can invade even well-managed natural habitats. Florida's latest scourge is the Old World climbing fern from Southeast Asia, which can blanket forest trees. When burned, the fern carries fire into tree canopies that are normally protected from fire by their height from the ground. The trees have fire-resistant bark, but a crown fire kills them.

Unfettered demons such as Brazilian pepper, paperbark, and the rest of Florida's plagues seem to have enormous niches with limits that are impossibly wide. By completely trouncing the competition, these plants spread into habitats where they would not normally be able to grow at home. In Queensland, Australia, for example, paperbark is confined to seasonally wet places, but in Florida it also invades areas with standing water and well-drained uplands. By comparison, alien plants that have been successfully controlled by the introduction of natural enemies have shriveled niches that are much more specialized.

A fascinating example is the perforate St. John's wort (or Klamath weed), an alien from Europe that was once a pest in the grasslands of California and that is now controlled by an introduced beetle. St. John's wort hasn't disappeared altogether, but is now only to be found in shaded places where once it enjoyed the sun. The reason is that the beetle that lays its eggs on the plant and whose grubs eat it avoids shade, so plants growing there find a refuge from this herbivore. The combination of natural enemies that defoliate the plant and other plants that compete with it have turned St. John's wort into a species that skulks in dark corners. Looking at its distribution in California today, who would dream that St. John's wort was once a demon of the pastures, not the shy and retiring plant it now appears to be? These changes were wrought by predation and competition acting together. Just like in grasslands (chapter 6), herbivory weakens the demon and competition then confines it to a narrow niche. How many native plants are similarly restricted we do not know, perhaps many.

It would be very useful to be able to predict which among the thousand or so nonindigenous plants growing wild in Florida that are presently harmless will one day become demonic. If this could be done, these species might be removed before they prove problematic. Unfortunately, as we already know, there is a demon in every plant and research has not yet discovered a reliable method of predicting which plants will realize this potential. The best predictor of trouble is a plant's record of prior offenses. If a species is already a demon in one area of introduction, it is highly likely to cause problems elsewhere too. The difficulty is that not every demon has a previous record of invasiveness. The Exotic Plant Pest Council of Florida keeps a list of species that have known invasion potential. Despite the recognized dangers, many of these "usual suspects" are still commercially available in Florida.

Each alien species has its own history and an individual demonology, but there is a common feature found in many plant invasions that offers a warning and an opportunity. In the histories of many invasive plants in Florida, there is a delay of several decades between its introduction and its eventual spread. The warning in this knowledge is that there must already be new demons in the pipeline. The opportunity is that delays may provide time to spot demons and to eradicate them before they get out of hand. Brazilian pepper trees took fifty years to become pervasive and paperbark was present for a similar period before a problem was recognized. Japanese honeysuckle was grown in gardens for eighty years before it ran amok from the Gulf of Mexico to Massachusetts, reaching as far west as Kansas. Delays of the kind seen in the spread of alien plants in Florida have also been observed elsewhere and may be an inherent feature of the invasion process. Perhaps a species must reach a critical density of colonies before its spread begins to take off, or maybe it takes time for natural selection to bring about adaptation to a new habitat.

There is only one case in Florida where the cause of the delay is really understood—that of the laurel fig. A native of Asia, this fig was grown as a street tree in Florida for many decades; it never set fruit, however, because it requires pollination by a particular species of tiny fig wasp, which had not been imported with the tree. Then, sometime in the 1970s, fig wasps must have arrived with a consignment of new plants, because laurel figs started to mature their fruit and to spread by seed. Fig seeds are dispersed by birds and the seedlings can grow in rock crevices, on buildings, and in the leaf-bases of palm trees. Laurel fig has now invaded many natural areas in southern Florida. Knowing that invasive plants tend to have a refractory period before their spread takes hold suggests that there is a window of opportunity when a small effort at control could avoid big problems later: a stitch in time saves nine. Research is needed on what happens at the transition between the refractory period and the period of spread, so that invasions by new species can be nipped in the bud.

Florida's troubles with alien invasive plants illustrate the practical need for a better understanding of what turns plants into demons abroad. Joseph Hooker summed up the problem rather well in 1853 in the book he wrote on returning from his travels in the Southern Hemisphere (chapter 1): "We have the apparent double anomaly, that Australia is better suited to some English plants than England is, and that some English plants are better suited to Australia than those Australian plants which have given way before English intruders."

To these anomalies might be added those Australian plants such as paperbark that grow better in Florida than do the natives of the Everglades. One might have thought that such anomalies would have worried Darwin because they strongly suggest that native plants are not as well adapted to home conditions as are species from halfway around the world, and that plants can grow better in foreign parts than in the region in which they evolved. How can natural selection explain this? (And, lest any creationist take comfort from this question, note that it is just as much of a problem if you believe that God placed each species in its right place.) Strangely, this is a question Charles Darwin ignored in the two chapters on geographical distribution in *The Origin of Species.*

Eighty years after Joseph Hooker puzzled over the anomalous behavior of traveling plants, another Englishman raised the same question. In the words of the arch snob and elegant wit Noel Coward: "Why, oh why, do the wrong people travel, when the right people stay at home?"

Is the explanation of demon behavior that the wrong plants travel, or that travel itself creates the problem? Since we can compare how the same species behave in indigenous and in alien populations, we do actually have an answer to this question, but it comes in two parts. The first has to do with the likelihood of a plant reaching foreign shores. Species with larger natural geographic ranges *are* more likely to turn up as aliens elsewhere. In this respect it is the plant that "elects" to travel. The second part of the question has to do with whether, once arrived, an alien plant then becomes an invasive demon or just sojourns awhile. Here, the answer is that there is no reliable correlation between a plant's abundance at home and how it behaves abroad. A good example is found in a group of three herbs in the genus *Impatiens* that occur as aliens in the British Isles. The most demonic of the three is Himalayan balsam (*Impatiens glandulifera*), but this species has the smallest natural geographic range, being confined to a narrow altitudinal belt between two and two-and-a-half thousand meters in a part of the western Himalaya. *Impatiens capensis*, the least widespread alien of its genus in Britain, has the widest natural geographic range of the three.

Evidently, travel itself turns some plants into demons. The ultimate proof of this is where two regions have swapped plants. The Everglade's gift to the home of paperbark is pond apple. In an irony of ironies, pond apple is now becoming scarce in Florida, but it has invaded Australia's paperbark swamps where it is considered the greatest of threats to the tropical wetlands in northern Queensland. Paperbark itself is a declining species in Queensland. Paperbark and pond apple each seem to grow better in the other's habitat than in their own. This is a severe shock to anyone who believes in the perfection of nature. What is going on?

The likeliest explanation is that travel enables demons to escape from the natural enemies such as insects, fungi, and diseases that attack them. The successes of biological control, which reunites plants with their enemies, bear this idea out. When we travel abroad we tend to catch local diseases or succumb to bugs in the water or the food. For plants it is quite the reverse. When they leave their native range they carry no baggage (not even leaves if they travel as seeds) and they usually leave behind all their natural enemies.

In Europe, purple loosestrife is plagued by a whole specialized fauna of herbivorous insects, but in North America these are absent and not one of the native insects has developed a taste for this exotic dish. Instead, not surprisingly, North American insects eat North American plants. This puts native plants at a disadvantage when they come into competition with aliens. The native wetland flora is weakened by attacks from its natural enemies, but the alien is not. This may well tip the balance into demonic behavior for a plant like purple loosestrife that is already so well prepared for this role. In sports competitions it is the home team that usually has an advantage over any visiting team. In the natural world it is often the reverse and visitors dominate the play.

The altered conditions that aliens encounter outside their natural range must also explain why, on occasion, they actually do better than either their confreres back home or the natives that are adapted to local conditions. Evolution is not a magic wand, and even though it can provide a plant with armaments against its natural enemies, it cannot banish those enemies entirely because they too evolve in response. By contrast, plants that, with human help, leap ocean barriers do get a magic carpet ride to a land of Cockayne where no disease will touch them and every herbivore finds them distasteful.

If escape from natural enemies explains the demonic behavior of alien plants, there still remains another mystery that is less easily explained. Why is the New World so troubled by plants from the Old one, while Europeans have much fewer problems with New World weeds? Darwin even pulled the leg of his American friend Asa Gray on the subject, asking him in a letter "Does it not hurt your Yankee pride, that we thrash you so confoundedly? I am sure Mrs. Gray will stick up for your own weeds. Ask her whether they are not more honest, downright good sort of weeds." To which her riposte was that American weeds were "modest, woodland, retiring things; and no match for the intrusive, pretentious, self-asserting foreigners."

Joseph Hooker described the situation as a "total want of reciprocity in migration." Trade and travel brought aliens plants from Europe to North America, but although human colonization was one-way, the traffic of goods and ships was not. Raw materials transported from all over the globe brought the seeds of alien plants to Britain in the nineteenth and early twentieth centuries. Imports of wool were a particularly rich source of seeds, amounting to 900 million tons a year just before World War II. Wool from Australia, Africa, southern Russia, Asia, and South America arrived at the woolen mills where it was washed and the seed-laden water was passed back into local rivers. Galashiels, which stands on the River Tweed in Scotland, just over the border from England, became a mecca for botanists seeking chance encounters with exotic plants. The wool supplied the manufacture of tweed cloth that was once in great demand.

In 1919, Ida Hayward and Claridge Druce published a book cataloging their discoveries made on the banks of the Tweed. They identified no fewer than 348 species of "wool alien." Nearly half were from Europe and the Near East, but significant numbers also came from Australia, South Africa, and the Americas. Now these plants are as rare in Galashiels as a tweed jacket on a Paris catwalk. The only survivor from the 348 species recorded by Hayward and Druce is a New Zealand plant of the rose family, the pirri-pirri bur. This plant found its way onto the Holy Island of Lindisfarne, which lies at the mouth of the River Tweed, where it is now a local nuisance. The Holy Island is famous for the illuminated Lindisfarne Gospels produced on the island in the eighth century. A translation of the Gospels from Latin into Old English was added between the lines of the manuscript in the tenth century, making Lindisfarne the source of the earliest surviving copy of the Gospels in the English language. The woolen industry of Galashiels might have been designed for the specific purpose of introducing troublesome plants into Scotland, so efficiently did it gather, concentrate, and propagate aliens along the banks of the Tweed. What an irony that all that now remains of the wool aliens is a castaway demon on Holy Island!

The fate of the wool aliens of Galashiels could not underline more strongly the "total want of reciprocity in migration" that Joseph Hooker described a century and a half ago. Only about 5 percent of the plant species found wild in Europe come from elsewhere, while the comparable figures for other regions of the Northern Hemisphere, such as Ontario in Canada or Florida in the United States, are about 30 percent. The wool aliens demonstrate that this difference cannot be due to a failure of the seeds of alien plants to reach Europe, and certainly Europeans are as keen to introduce and grow plants from foreign places as gardeners anywhere are. Hundreds of thousands of varieties and species of plants are grown in European gardens and have only to leap a fence to run wild. Surprisingly, few are successful. Why is this?

Climate must be part of the answer. Plants that can tolerate the climate of northern Europe when tended by a gardener cannot do so when exposed to the wild, where they also have to contend with competition. As always, Darwin perceived the situation clearly. He remarked that there were a "prodigious number of plants in our gardens which can perfectly well endure our climate, but which never become naturalised, for they cannot compete with our native plants, nor resist destruction by our native animals."

The cooler the climate, the harder it is for aliens. There are many more alien plants in the south of England than in the chillier north or in Scotland, and the same pattern is repeated on a larger scale throughout Europe. However, an inhospitable climate cannot be the whole reason for the different success of aliens in the Old World and the New. Even in warmer parts of Europe, few aliens turn demonic. A dozen of the pasture weeds that infest the grasslands of northern North America, including St. John's wort, are native in British grasslands, but not one alien grassland species has managed to reciprocate. Climate cannot explain this imbalance, so what other explanation can there be?

The North American grasslands that have been invaded by aliens are not those few remnants of native prairie where once the bison roamed. The grasslands that alien weeds invade are as foreign to North America as the aliens themselves and they are grazed by livestock whose ancestors also came from Europe. In fact, these grasslands are whole transplanted ecosystems. The grassland plants from Europe that are palatable to livestock have acquired a Green Card and obtained legitimate employment feeding cattle and horses. Some have even acquired an American alias such as "Kentucky bluegrass," which back home in England goes by the name of "smooth-stalked meadow grass." The disowned, invasive weeds are simply their less palatable traveling companions.

These North American grassland communities were created, wittingly or otherwise, by European settlers raising European animals. It should therefore be no surprise that European plants found the conditions created by this kind of farming, which was unknown in North America before European settlement, so much to their liking. The introduction of European livestock into North America, Argentina, Australia, New Zealand, and South Africa quite literally prepared the ground for the invasion of alien plants from the same source. Realizing this, and looking back across the ocean, it suddenly seems less surprising that Mrs. Gray's "modest, woodland, retiring" American weeds have failed to invade Europe. Were North American woodland environments to be recreated in Europe, the story might be quite different.

Alien plant invasions are very rarely accidents of nature. Some such as kudzu or paperbark were deliberately aided by misguided human action. Others, like the aquatic plants released into Florida's waters to provide commercial wholesalers with a convenient source of aquarium plants, were accidents waiting to happen. Another group are pasture weeds on a spree in habitats that feel to them just like home, but without their accustomed pests and diseases. There are many flavors of folly, and now that we are familiar with some of them it would be wise to avoid past mistakes. This must mean looking with a wary eye at any possible source of new demonic plants.

9

New Demons?

Think of your favorite plant. Is it a rose with carmine velvet petals and a heavenly scent? Perhaps it is a graceful tree, or a luscious fruit like a perfectly ripened strawberry or mango, or the humble potato that makes wonderful French fries. Everything a plant can be, and everything it can do, is written in its genes—turning carbon dioxide and water into sugars by the power of sunlight; defending itself from enemies; the color, scent, and form of flowers; the growth and shape of leaves; and, most important of all, the production of seeds and pollen that transmit the genes to future generations.

It is a safe bet that your favorite plant is highly bred. Taking wild species with promising characteristics of form, scent, or flavor, plant breeders have "improved" on nature by selectively breeding individuals that show desired characteristics in their most extreme form. To obtain larger blooms, roses with the biggest flowers are crossed with one another and this is repeated with the biggest of their progeny. The result of this process, repeated generation after generation, is a rose with flowers much bigger than anything ever found in the wild. Darwin argued that the astonishing feats of transformation that artificial selection could bring about in domestic breeds of animals and plants was powerful evidence that natural selection, if given enough time, could achieve even greater wonders.

Consider the cabbage. This vegetable is a domesticate of the wild species *Brassica oleracea* found in coastal areas of Europe. The wild plant is an inedible-looking thing with a tough stem supporting only a few, bitter-tasting leaves. There is no archaeological record of cabbage domestication, but Theophrastus (ca. 372–286 B.C.), a pupil of Aristotle with a particular interest in plants, mentions three kinds in his writings, none of them with the compact heart of the modern vegetable. Over the centuries, artificial selection has taken this unpromising starting material and produced a plethora of vegetables.

In modern cabbages, the terminal bud of the plant is massively enlarged to form a heart; in kale, the stem leaves have proliferated and in many varieties are curled or attractively colored in silver and purple; in Brussels sprouts, the side buds on the stem have become numerous and tightly compacted; in cauliflower, the immature, terminal flower head is enormously swollen, while in broccoli the side branches carry smaller heads of edible flowers; finally, in kohlrabi, artificial selection has produced a swollen stem base that looks like a turnip with leaves sprouting from its sides. All these once-cabbages have also had most of the bitter taste of their wild ancestor bred out of them. Just enough remains to give them flavor.

The full scientific names of wild plants and animals carry a postscript indicating which botanist first described them. The "L" after *Brassica oleracea* L. tells us that Linnaeus himself named the wild cabbage. However, the geniuses of plant breeding who spent generations creating an entire crew of vegetables from *B. oleracea* L. are nameless and unrecorded, their only memorial the plants themselves. The farmers and gardeners who, since the dawn of agriculture, have conjured comestible plants from wild species had never heard of genes, but they were the earliest geneticists. Through artificial selection, they unwittingly manipulated genes into combinations that brought out the desired characteristics of plants and animals. At times they would add the genes of another species to the stew by hybridization. For example, the modern varieties of rose that go on flowering time after time throughout the season all owe this ability to the genes of *Rosa chinensis* var. *spontanea*. Modern bread wheat contains the genes of three different wheat species.

Genes are instructions written in a molecular code. If sequencing a gene is the equivalent of reading these instructions (chapter 2), then genetic engineering is the equivalent of rewriting them. No longer need plant breeding rely on the unwitting manipulation of genes. They can be identified, read, and copied from one organism and inserted into another. The science of genetics has come of age. Imagine the power to modify a plant's genes so that it, and all its offspring, grow better; look, smell, and taste better; last longer; or produce new proteins and medicines. Imagine vaccines or anti-AIDS drugs as cheap as soybeans. This is the promise of genetic engineering, but what is the reality? Can we really design nature to our own ends, or are we being tempted by a Faustian bargain to trade our most precious possessions for illusory reward? Will genetically modified plants released into the environment become the new demons?

The genetic modification of plants is not new. What is new is the technology for transferring genes highly selectively and between organisms as distantly related as animals, plants, and bacteria. Nature's way of transferring genes between related species is through hybridization, but though hybrids may be viable and are sometimes vigorous, they are often sterile. The mule, a hybrid between a donkey and a horse, is the classic example in the animal kingdom. The problem for such hybrids is that when their reproductive cells come to divide to produce eggs and sperm they must first go through a process in which the chromosomes derived from the hybrid's mother pair up with those from its father. When mother and father belong to the same or to very closely related species, pairing proceeds normally; but when a hybrid's parents are more distantly related, pairing fails. In animals there seems to be no way around this problem and hybrids are usually evolutionary dead ends. Mules must be made anew each generation by mating a horse and a donkey. The notoriously stubborn mule is an evolutionary full stop. The situation in hybrid plants, however, is quite different.

In 1899, gardeners at Kew discovered something unusual in a seedling growing from seeds collected from the primrose *Primula floribunda*. It had leaves the shape of its mother's, but it otherwise resembled another species, *Primula verticillata*. Experimental crosses between the two species were made that produced intermediates resembling the hybrid, thus confirming its probable origin. The hybrid had eighteen chromosomes, the same number as its parents, and was sterile. However, it grew vigorously and was vegetatively propagated and widely distributed to gardeners under the name *Primula* X *Kewensis* (the "X" denoting a hybrid origin).

The Kew primrose did not remain a mule for long and in 1905 a plant set good seed in a nurseryman's garden; other plants would do the same at Kew in 1923 and at a research institute in 1926. In each case it turned out that these fertile plants had thirty-six chromosomes, twice the number of the original hybrid and its parents, and they bred true. Evolution had overcome the block to fertility by simply doubling the chromosomes from each parent to produce a polyploid. Since the polyploid had a double set of chromosomes from each parent, every chromosome had a compatible partner. Though restoring fertility to the hybrid, chromosome doubling prevented interbreeding with the parents of the hybrid and thus evolution had produced a new species, isolated from any further exchange of genes with its progenitors.

Polyploidization, or multiplying the original chromosome number, seems to have occurred in at least half of all flowering plants, some having multiplied their chromosome number more than once during their evolutionary history. Certain whole groups such as willows and poplars are ancient polyploids. Not every polyploid has a hybrid origin, though the majority probably do, and so the transfer of genes between related species cannot be considered unnatural.

If natural gene exchange between species is so widespread in the evolution of flowering plants as to be present somewhere in the majority of lineages, do we have anything to fear from artificial genetic modification? Need we worry about something so natural? There are actually two questions here. The first is whether a natural process may contain dangers of its own and the second is whether artificial genetic modification is equivalent to the natural process. If it is clearly different from the natural process, are its own dangers greater or smaller and how can we tell?

The existence of Darwinian demons should be evidence enough to convince us that what is natural is not necessarily risk-free. If new species did not, at least initially, possess extraordinary powers of multiplication and spread it is unlikely they would ever become established. That so many plant species apparently evolved through hybridization means that genetic modification may often endow a species with those very demonic powers that a new species needs to be successful. Several cases of exactly this have been observed.

Species of cord grass are found in coastal plant communities on both sides of the Atlantic. Toward the end of the nineteenth century, a North American species found its way across the ocean, probably in a ship's ballast, to Southampton Water in southern England. There it grew and crossed with the native species and established a population of sterile hybrids that were first recorded by botanists in 1872. In a pattern that has now become familiar, a fertile, polyploid version of the hybrid was found in 1892—the union of transatlantic cousins had given birth to a new species that was subsequently named *Spartina anglica*. The new species was able to invade mud flats that were inhospitable to its native parent, and over the subsequent century *S. anglica* spread around the coasts of Britain, displacing more diverse plant communities and reducing the area of tidal mudflats used for feeding by wading birds. One of the nature reserves where *S. anglica* is now found is on the Holy Island of Lindisfarne (chapter 8), where it is encroaching on mudflats used by Brent geese and widgeon.

The ability of *Spartina anglica* to colonize and bring stability to mudflats is deemed useful in some parts of the world, and the species has been introduced for this purpose as far afield as China. Its American parent has been transported for the same reason from its native east coast across the United States to San Francisco Bay, where it has begun to form hybrids with and displace the native west coast species *S. foliosa*. Is there another invasive species of *Spartina* in the making?

The parents of hybrids need not come from distantly separated or alien populations. There are several hybrid species of recent origin among native sunflowers in the United States. The common and the prairie sunflower are two widespread species whose geographical ranges in the western United States largely overlap. They prefer different soil types, the common species being restricted to heavy soils and the prairie sunflower to dry, sandy soils, though swarms of hybrids are common where the two species grow near one another. The hybrids are often sterile, but a handful of populations have been found in Texas that contain plants that resemble hybrids, but that are fertile. When first discovered, opinion among botanists was divided as to whether these populations really belonged to a new species or not. Those who believed they did named the new sunflower *Helianthus paradoxus*, and recent genetic analysis has vindicated them.

The genes of *H. paradoxus* turn out to be a mixture of genes from its two parent species, but evolution has rearranged them in such a way as to prevent the new species producing fertile seeds by mating with either of its progenitors. This incompatibility isolates *H. paradoxus* from its two parents and the new species thus qualifies as a separate entity. Although the genome of the *H. paradoxus* appears to be largely a reshuffling of its parent's genes, it occupies a habitat that neither parent can colonize—*H. paradoxus* is found in brackish waters that have a salinity up to one-third that of normal seawater. This paradoxical sunflower seems to be another example of offspring entering territory beyond the ken of their parents and of new species requiring new niches. That new hybrid species like *Spartina anglica* and *Helianthus paradoxus* occupy new niches confirms Darwin's prediction about the evolution of diversity (chapter 3). Hybridization is the third important mechanism by which new plant species arise and is of similar importance to adaptive radiation and geographical isolation in the evolution of plant diversity. (It is much less important in the animal kingdom).

These examples of hybrid speciation are perhaps in danger of giving the impression that hybrids generally succeed, but they do not. Hybridization has been more thoroughly cataloged in the native British flora than anywhere else, and though hundreds of natural hybrids have been observed, successes seem to be a comparative rarity. However, though a million hybrids may fail, natural selection can seize on the rare exception and catapult it into a new habitat or a new region. Repeated hybridization is not needed for this to happen. An occasional crossing between species somewhere in a lineage may be enough to introduce the genes required, and these can spread through a population without hybridization being repeated. Genetic boundaries between species need only be slightly leaky for this process to happen. Can new genes acquired by the botanical equivalent of a one-night stand impel existing species into becoming demons?

King Louis XIV of France, known as the "Sun King" for the golden radiance of his court, was renowned for his style and extravagance. His botanical legacy included more than thirteen hundred new species discovered by Joseph Pitton de Tournefort, professor of botany at the Jardin du Roi in Paris, whom Louis XIV ordered to explore the eastern Mediterranean. Among de Tournefort's finds, first seen near the Turkish town of Pontus, was *Rhododendron ponticum*. This evergreen shrub appears to have first been brought to Britain from southwestern Spain in 1763, and it was welcomed into all the best gardens. After a century of garden ease and with the enthusiastic help of Victorian gardeners who liberally planted *R. ponticum* as a rootstock on which to graft less vigorous and more exotic *Rhododendron* species, the plant went wild. Rootstocks suckered and spread and the species' tiny seeds helped it travel the length and breadth of Britain, spreading along railway embankments and taking up residence in woodlands.

Only two things appeared at first to limit this demon: it could not abide lime-containing soils nor survive very cold winters. *R. ponticum* continues not to pose a problem on calcareous soils, but it is now naturalized up into the far north of Britain: the demon appears to have evolved cold tolerance. Recent research has discovered that populations of *R. ponticum* in eastern Scotland, which is the coldest region of Britain, contain genes that come from a more cold-tolerant species, *Rhododendron catawbiense.* This species was an early horticultural introduction from North America and can tolerate temperatures much colder than anything ever recorded anywhere in Britain. Somewhere in its history of colonization and spread through Britain, and possibly on more than one occasion, *R. ponticum* evidently hybridized with *R. catawbiense* and seems to have acquired genes for cold tolerance from this source. Cold temperatures, possibly aided by gardeners selectively propagating their hardiest stock, seem to have selected those hybrid rhododendrons able to survive in eastern Scotland, thereby pushing the invasion outward to farther limits.

All plants, even demons, have their geographical limits, and there is no doubt that a plant's genes play an important role in determining where those limits lie. The case of *Rhododendron ponticum* indicates that new genes, acquired in this instance by hybridization, can help a species spread, but there is no suggestion that *R. ponticum* was harmless before it acquired them. So far as we know, the shrub had already rampaged through most of Britain without any help from hybridization. The same appears to be true in another case in Florida. Two species of *Casuarina* are among the invasive Australians found in Florida, but they also hybridize to produce a third demon that combines the tall stature of one parent with the salt tolerance of the other.

What, if anything, do these examples tell us about the likely environmental dangers posed by genetically modified plants? First, problems created by invasive alien species tell us one thing very clearly: invasions do happen and can be serious. We ought to take all precautions to avoid creating new demons. Second, they tell us that genes moved between species, by hybridization in these cases, can cause problems by helping plants to invade new habitats. Note, however, that moving species around seems to have created far more trouble than moving genes between species. Nonetheless, this does seem to add up to a case worth investigating. The next step is therefore to look at what we know about the ecology and genetics of transgenic crops themselves.

The first questions ask how similar transgenic plants are to natural hybrids, and which genes are transferred and what their effects on the recipients are. The answer to the first question is simple. Transgenic plants (transgenes, for short) are as like natural hybrids as are the plants produced by conventional plant breeding. The reason is that transgenes are conventional plant varieties into which a small number of genes have been transferred. Conventional plant breeding combines by hybridization entire genomes containing thousands of genes, and then selects among the plants that result.

Does the fact that only a few genes are transferred by genetic engineering make transgenes safe? Genetically modified (GM) corn, potatoes, cotton, and soybeans are widely grown in the United States without any particular safeguards being required on the grounds that they are "substantially equivalent" to the conventional varieties that were modified to produce them. However, the concept of substantial equivalence is a major bone of contention between the supporters of GM technology and its opponents. If a new GM crop is substantially equivalent to an existing variety, how can it simultaneously be novel enough to be patented? Nonetheless, the biotech companies such as Monsanto that own the patents on these plants have managed to get away with both claims, at least in North America.

Sceptics of genetic modification are mainly worried about three issues: potential threats to human health from consuming food from GM crops; the possibility that GM crops might have adverse environmental effects, including the risk that some might become invasive weeds; and contamination of non-GM crops by genes carried in pollen or seeds. An important issue, about which there is often confusion, is whether alleged adverse health or environmental effects derive from the GM technology itself or from the genes that the technology has been used to transfer. Supporters of GM tend to argue that the technology is just a better way of achieving what plant breeders have done for millennia and what has happened in nature throughout the history of life.

At a debate on these issues that took place in Atlanta, Georgia in December 2001, Toby Bradshaw, a plant biotechnologist who works on GM trees at the University of Washington, said "It's a myth to think that humans invented the transfer of genes into plants. It has been done for millions and millions of years by soil bacteria, and it is possible to convince those soil bacteria to transfer any bit of DNA that you would like moved into a plant." Earlier the same year, a group calling itself the "Earth Liberation Front" had burned Bradshaw's lab to the ground because they claimed that his work "continues to unleash mutant genes into the environment that (are) certain to cause irreversible harm to the forest." Nothing is certain about either the harmful or the beneficial consequences of GM—that is a major part of the problem. If anything were certain, it would be easier to decide the issue.

However, the genes that Toby Bradshaw and others trick soil bacteria into transferring into plants are not the genes that these bacteria naturally transfer. Even if the process is natural, the product is not. Proponents of GM focus on the intrinsic safety of the tools for gene transfer and tend to consider that problems, if they exist, are only associated with the effects of particular genes. In other words, it is not the GM tool, but what you do with it that matters. Sceptics tend to ignore this distinction, perhaps taking the view that exotic genes couldn't be transferred and become a problem without GM technology.

The counterargument to this is that conventional plant breeding sometimes produces dangerous products such as poisonous potatoes or allergenic celery, but these hazards receive no attention because the products don't get to market. Conventional breeding has also produced herbicide resistance in some crops, a trait that is commonly engineered into transgenes. These examples support the view that there is nothing uniquely risky about GM *per se.*

The suspicion that GM crops are inherently different and dangerous was aroused early on in their development by the presence of a gene for resistance to antibiotics that was incorporated into many of the first generation of transgenic crops. This gene was there as a byproduct of the genetic engineering process, not because it served any useful function in the crop itself. The idea that food might contain something that inactivates an antibiotic, and the possibility that a gene for this substance might spread from GM crops, caused understandable public alarm, particularly in Europe. Newer GM technology can dispense with antibiotic resistance genes, but the public perception of GM food is not so easily altered.

There is almost no limit to the kinds of genes that can be inserted into plants, but at the time of this writing two types are attracting most attention: genes for herbicide tolerance and a gene that produces a toxin that kills caterpillars. All farmers fight a continual battle against weeds and insect pests, so crops, chemicals, or farming systems that can reduce the damage done by these two scourges would help them, *if* they can sell the crop. The "if" is crucial: the public in Europe and Japan, and increasingly in North America, is suspicious of GM, and many supermarket chains and food processors refuse to buy the crops.

From the farmers point of view, the perfect herbicide is one that kills weeds without harming the crop. Herbicide manufacturers like Monsanto, which produces Roundup (glyphosate), have produced GM crop varieties that are tolerant of their proprietary brand herbicides. By sowing proprietary GM seeds, also supplied by Monsanto, and applying the proprietary herbicide, a farmer can in theory grow a weed-free crop. By the year 2000, millions of hectares of GM canola were being grown this way in western Canada, consisting of three main varieties, each one resistant to a different herbicide.

Canola is an unusual crop because it is recently domesticated and still has the vestiges of two characteristics found in its wild relatives: seed shattering and seed dormancy. Shattering means that some seeds are shed before harvest, while dormancy allows some of them to survive several years in the soil. Because it has these properties of a wild plant, canola often appears as a "volunteer" weed in follow-on crops. Volunteers can be controlled with a herbicide unless, of course, they come from a GM crop, in which case they are tolerant of the proprietary brand.

In spring, canola fields are a solid carpet of yellow flowers, testimony to the fact that the plant is insect pollinated. Insects and wind carry canola pollen from flower to flower and from field to field. It hould have been no real surprise, then, when reports began to come in that canola volunteers tolerant of more than one herbicide were showing up in Canadian fields. The combination through cross-pollination of genes from different GM varieties is known as "gene stacking" and is a worrying development that threatens to produce a serious weed problem in the prairie provinces of Canada. Gene stacking is turning canola into a demon weed that many of the common herbicides cannot kill.

Weeds of arable fields are nothing new and farmers may learn to live with gene-stacked herbicide tolerant canola, though at a price. However, gene stacking robs agribusiness of the notion that GM puts them in full control of the crop. At the very least, it shows an extraordinary indifference on the part of the companies that developed the plants to what would happen to GM crops once released. No competent plant scientist could claim ignorance of the possibilities when adjacent populations of insect-pollinated, herbicide-tolerant plants came under selection from several herbicides. We known full well that under strong selection, evolution responds with tolerance because this has happened many times already. Herbicides are used so widely and so heavily that over 150 weed species worldwide have evolved tolerance to at least one of the chemicals, and new cases are being found all the time. Herbicide tolerance could potentially spread from canola to wild radish and several other related wild species that grow in Canada and elsewhere.

Gene-stacked, herbicide-tolerant canola looks like a piece of folly out of the same recipe book as the seeding of the Everglades with paperbark (chapter 8). However, an important difference, and a straw that might be grasped in defense of GM, is that canola and other GM crops do not seem to be an immediate threat to *natural* habitats in the way that invasives such as paperbark often are. Experiments in Britain in which conventional and GM varieties of canola, corn, beet, and potato were sown into natural habitats found that only potato populations survived more than four years. In no case did GM varieties survive significantly better in the wild than conventional ones. The plants used in these experiments carried herbicide-tolerant or insect-resistant genes, but transgenic plants carrying other genes, for example for drought tolerance, might behave quite differently. Likewise, forest trees or other plants that are not as domesticated as these arable crops might be more likely to become invasive. China is at the forefront of GM technology and one project being worked on there is salt-tolerant Chinese water spinach. This plant is already a problem in many parts of the world (chapter 8), and salt tolerance could allow it to invade brackish water, widening its scope for damage considerably.

A crop grown very widely in the United States is GM corn containing a gene from the bacterium *Bacillus thuringiensis*, or *Bt* for short. This bacterium is a natural enemy of leaf-eating caterpillars and some bacterial strains kill other insects, including mosquitoes. The bacterium can be cultured artificially and is permitted for use in organic agriculture. *Bt* corn, so-called, contains the bacterial gene that makes the natural caterpillar-killing toxin. The toxin defends *Bt* corn plants against caterpillars of the European corn borer and other serious pests. In theory, a built-in insecticide like the one produced in *Bt* corn ought to bring environmental benefits because it avoids the need to spray insecticides that can harm nontarget insects as well as pests. Surely, then, isn't this a case of a truly beneficial GM technology at work? The promise is there, but so are some dangers.

One danger is that the *Bt* gene will escape from crops into wild relatives, which will then become demon weeds. Where crops and wild relatives grow near each other, which happens surprisingly often, some cross-pollination is inevitable. Experimental transfer of *Bt* from crop sunflowers to wild relatives by cross-pollination has shown that in wild sunflowers, the *Bt* gene can greatly reduce the damage done to plants by caterpillars and increase seed production by 50 percent or more. Whether this protection from some of their natural enemies is sufficient to turn wild sunflowers into demons is not known, but the possibility must exist.

A possible risk of another kind was reported in 1999 by three entomologists from Cornell University. In some preliminary laboratory experiments they found that monarch butterfly caterpillars feeding on leaves that had been dusted with *Bt* corn pollen grew less well than caterpillars feeding on leaves dusted with pollen from non-*Bt* corn. Monarch butterflies are an icon of nature conservation in North America, making migratory flights the length of the entire continent from Canada to Mexico. Any suggestion of a threat to their survival could be expected to provoke a strong reaction, and it did. Environmentalists seized on the research as proof that GM corn was environmentally harmful and held public demonstrations against it dressed as monarch butterflies.

The scientific reaction was more sceptical because laboratory results often bear little relation to what happens in the field. The problem was, no one had published the results of an equivalent field experiment, so the report, preliminary though it was, was all that there was to go on. Field research was quickly commissioned and when the results came in two years later, it showed that harmful effects of *Bt* corn pollen on monarchs were negligible because exposure of caterpillars to the toxin around fields was very low. In any case, the *Bt* corn variety with the most toxic pollen has now been withdrawn from the market.

There are lessons for both sides in these events. On the side of biotech companies and the agencies that regulate them, the lesson is that environmental effects must be fully assessed over many years. Secondary and indirect effects due to ecological and evolutionary interactions between the transgene and other organisms must be experimentally investigated in the field. The implications for those of us who care passionately about the environment are just as important. We should recognize that properly regulated releases of GM crops could bring environmental benefits. The choice may not be just GM or no GM, but environmentally friendly GM versus environmentally damaging sprays. Growing GM cotton containing the *Bt* gene has enabled small farmers in China to significantly reduce their use of pesticides and to lower their costs. There also appear to be health benefits, with many fewer cases of poisoning among Chinese farmers growing *Bt* cotton as compared to those growing non-GM varieties that require more pesticides. If you accept that properly regulated and well-designed GM can be environmentally friendly, then you also have to accept that transgenes have to be studied in the natural environment to test that they are safe.

A large-scale attempt to test herbicide-tolerant GM crops for their effects on the biodiversity of weeds and insects was made on over sixty British farms between 2000 and 2002. The studies were at the farm scale, in recognition of the fact that the environmental impact of GM crops is likely to depend, at least in part, on how they are cultivated by farmers (for example, how often herbicides are used), not simply on the crops' genetic characteristics. In the event, major differences in farm management practice between GM herbicide-tolerant crops and non-GM controls were the decisive factor in how biodiversity was affected. Compared to non-GM controls, farm practices lowered weed and insect diversity in GM canola and beet, but apparently raised it in crops of GM corn. There was some question as to whether the beneficial effects in maize would be long-lasting, but on the basis of the experimental results the British government lifted its ban on the planting of GM corn, while maintaining a ban on the two other crops. Not long after the ban on GM corn was lifted, Bayer, the company marketing it, announced that it would no longer sell GM seed in Britain because it was not commercially viable. Nobody wanted it.

The British farm-scale evaluations are an object lesson in the complexity of the GM issue, and they illustrate how carefully such trials have to be interpreted. Though, in at least two cases, biodiversity in fields growing herbicide-tolerant crops was lowered, this does not demonstrate that GM technology itself, rather than herbicide tolerance, is the problem. The only way to test whether the technology, as distinct from the particular genes transferred, poses a risk would be to compare the effects of two herbicide-tolerant crops, one that had been subject to genetic engineering and the other of which had been made herbicide tolerant by traditional plant breeding. This experiment would be quite feasible but will probably never be done because it is already fairly clear that it is the properties of GM plants, not their origins, that determine whether they are an environmental threat or not.

The distinction between safety testing GM technology itself and testing GM products will probably strike anyone who has made up their minds about the dangerousness of GM as nit-picking in the extreme, but it is important. If the technology is not inherently dangerous, it could be used to the advantage of humankind and the environment. Before we throw the baby out with the bathwater, we ought to take a closer look at what the baby might grow into, if reared with caution.

In the realm of possible environmental benefits from transgenic plants, consider, for example, whether you would like to see elms, now extirpated from much of Europe and North America by Dutch elm disease, growing again in our countryside. Trees genetically modified to resist the elm disease fungus could be the best hope of restoring elm species. Likewise, the American chestnut, which was all but wiped out by chestnut blight in the first few decades of the twentieth century, might be restored from genetically modified, blight-resistant plants. To some, such as the American Chestnut Foundation, this is an exciting, if distant, prospect, though to others the genetically engineered resurrection of a tree that once dominated the broad-leaved forests of eastern North America might be the realization of their worst fears of GM.

As another example, think what progress might be made in genetically engineering crops suitable for sustainable agriculture—an agriculture that requires fewer pesticides and less fertilizer, for instance. To be sure, such an agriculture can be GM-free, but at what cost? Copper sulphate is permitted for use in organic agriculture to control fungal diseases, but it does not degrade in soil and becomes a pollutant. Might not fungus-resistant GM crops be safer for the environment and better for the farmer as well as the consumer? Is it rational to ban GM whilst allowing the use of copper sulphate in organic agriculture?

GM crops needs to be seen as part of a bigger picture. It's not just about herbicide resistance or *Bt* corn, but about what kind of agriculture we want. The GM crops that are the main cause of controversy today have been produced by large corporations that make their profits from industrialized agriculture. However, most breeding of new plant varieties is not done by these corporations, but by national and international agencies that are trying to feed people, particularly in the developing world. Sustainable agriculture is, or should be, their goal. Some of these not-for-profit organizations in countries where starvation stares them in the face are gung-ho for GM, and for good reason. Furthermore, if there is to be room left for wild plants and for nature in a future world with 9 billion people, we had better get the most we can out of our existing agricultural land, or our impact on the remaining wild places will spell the end of Eden.

10

The End of Eden?

Lorenzo Martínez stands proudly in his forest garden. He is a Popolucan Indian, whose ancestors have farmed in the Mexican rainforest since time immemorial. Plants are Lorenzo's livelihood and he knows them intimately—their needs, their flavors, and their uses. His garden, or *huerto*, is in a forest clearing and looks unkempt and disorganized to eyes that value tidiness and geometrical order. But the appearance of disorder in Lorenzo's garden is deceptive because it is organized along nature's lines and these do not circumscribe regimented beds. To see the natural order that is present here you need the eye of ecological knowledge.

Many of the trees in this garden are forest natives that Lorenzo has preserved for a purpose:

> We use the wood of this tree for tables. . . . This condoria has edible fruit which we can sell and the wood is also used for our houses. That custard apple has a very tasty fruit; it also comes from the forest. It's a tree that grows high and likes plenty of sun. This is a big lime tree used for jam, it grows in open conditions—doesn't need shade. This spearmint is put here because it would dry out in the dry season—it needs a lot of water. And these leaves are a pepper vine that we use to spice our tamales—it needs to grow in the sun and needs lots of water too. That's why we've got it here.

The *huerto* has different layers, like a natural forest. Lorenzo grows coffee bushes under the shade of trees such as the legume *Gliricidia sepium*. Nearby, in the ground layer, grow a tomato plant or two, and in the shade of an orange tree are some chilis. " Up-there," Lorenzo points into the branches of the orange tree, " is a pitaya cactus. It has seeded itself there. It produces a delicious fruit."

The fertility of the soil in Lorenzo's *huerto* is maintained by recycling household waste and the manure from chickens, turkeys, and pigs. Growing legumes helps maintain soil fertility too. These plants, such as *Gliricidia* trees and of course the ubiquitous Mexican beans, have nodules on their roots that contain bacteria which convert atmospheric nitrogen into a soluble form that plants can absorb. The *huerto* occupies a permanent plot, but Lorenzo also grows maize and beans in small, temporary fields called *milpas.* This is shifting cultivation in which a small area of forest is burned and then planted. The fire releases plant nutrients that support the growth of the first crop of maize. When this crop is over, the dead maize stalks are left in the ground to support beans which climb up them. After a couple of years the *milpa* is abandoned and it is recolonized by forest plants that help the soil recover its fertility.

The Popolucan Indians grow and use about 250 different plants in their *huertos* and *milpas*, but their form of agriculture, which has sustained them for millennia, has almost disappeared as the tropical forests of southeast Mexico have been felled and replaced by fields dedicated to raising a single cash crop such as pineapple, sugar cane, tobacco, or cattle. These monoculture crops spell an end to botanical diversity, to low-input agriculture, and to a kind of Eden.

Felix Castellaños is a farmer who knew the old ways and has experienced the new. " This was like a different country," he says, remembering a time only fifteen years before. " Then, we produced more with less work. And today, we produce less, and we need to put a lot more money and work into the soil." Pests were less of a problem in the old *milpas* and yields were higher too. " Before, we sowed one kilogram of beans and we produced one hundred kilograms; now you sow fifty kilograms and if you lift twenty more, that's a lot!"

Just as signs of disease can be detected in a patient's urine, ecological maladies now show up in the waters that drain the land. A stream in an undisturbed forest runs crystal clear, but rivers like the Coatzacoalcos that drain this area are laden with silt when they reach the sea, evidence of deforestation and soil erosion upstream. From this and other signs, Rodolfo Dirzo, a tropical biologist from the University of Mexico who has worked in the Los Tuxtlas area for many years, was sure that deforestation was getting worse, but he lacked the data to prove it.

Alarming estimates of some very high rates of tropical deforestation had been published by environmentalists, but they had been challenged by others who accused them of lacking hard data and making exaggerated and misleading claims. So, in 1989, Rodolofo

Dirzo and a colleague obtained aerial and satellite photographs that showed the extent of forest in the region as it had been in 1967, 1976, and 1986. A comparison of the three photographs showed that forest in Los Tuxtlas was being lost at four times the rate previously estimated for Mexican tropical forest. By 1967 the area had already lost 64 percent of its original forest cover, but over the succeeding twenty years what was left declined at a rate of more than 4 percent a year. By 1986 only 16 percent of the original forest area remained. Most of the forest has now become low-grade cattle pasture supporting only one cow per hectare.

To look at the three maps of forest cover in Los Tuxtlas that Rodolfo published is profoundly depressing. The one for 1967 shows a solid black, irregularly shaped blotch of forest occupying the center of the map. Imagine a close-up of the hide of a Friesian cow and you will get the picture. Only nine years later and the map shows the forest becoming fragmented, broken into tenuously connected patches so that the map now resembles the skin of a leopard more than the hide of a Friesian. Ten years more and we are looking at the coat of a Dalmatian dog. There are now spots of remnant forest dotted over the map, with a black core of forest remaining only on mountain slopes and in other inaccessible places.

Mexico is a megadiversity country. No one knows for sure how many plant species there are, but twenty thousand is a conservative estimate—nearly 5 percent of the entire flora of the Earth. Half of these plants are found nowhere else, and we owe to Mexico a cornucopia of crops including maize, beans, cocoa, squashes, and chilis, as well as garden plants such as dahlias, salvias, poinsettias, and countless cacti, palms, bromeliads, and orchids. The civilizations of the Maya, the Olmecs, and the Aztecs were all built on productive botanical foundations.

It is its varied climate and geography that has blessed Mexico with such botanical riches. Nearly every kind of ecosystem found on Earth is represented, from deserts to alpine meadows. The mountains of Mexico are clothed with every variety of forest, from tropical rainforests in the foothills to cloud forests where tropical and temperate species meet, to pine forests near their summits. Even the diversity of some temperate plants has blossomed in unaccustomed variety in this terrain where there are forty species of pine and nearly four times that many of oak.

Though they may receive just a few millimeters of rain each year, the deserts of Chihuahua and Sonora in northern Mexico are filled with cacti and other drought-tolerant plants that give the lie to the notion that deserts are deserted. At the other extreme, the tropical rainforests of southeast Mexico receive five meters of rain each year. In common with other tropical forests, those of Mexico are so diverse that many of their botanical riches are still unknown to science. As recently as the late 1980s, a plant so strange that it was placed in a new family all of its own was discovered growing in the Lacandon jungle of southern Mexico. *Lacandonia schismatica*, as the new species was named, is the only plant known to have its male organs in the center of the flower, with the female parts surrounding them. The sexual organs in its flowers are inside-out. Today the Lacandon jungle, like Mexico's other forests, is being fragmented by deforestation; an analysis of aerial photographs for the region by Rodolfo Dirzo suggests that as many as 22 percent of its plant species could be on an irreversible path to extinction by 2035.

The loss of natural habitats and the plant species that live in them is a global phenomenon. When I first saw Rodolfo's maps of Mexican deforestation, I was forcibly reminded of the strikingly similar pictures depicting the decline of heathland habitat in the county of Dorset in England. This vegetation type—dominated by heathers and growing on poor, sandy soils—is an English speciality. Though not very rich in plant species, it is home to rare animals such as the Dartford warbler, sand lizards, and smooth snakes. The heathland landscape is the dark, brooding milieu for Thomas Hardy's novel *Return of the Native.* In the 1840s, when the novel is set, the map of Dorset heathland displayed a Friesian pattern; by the end of that century, when the novel was actually published, Hardy observed that what had been uniform heath in 1840 "is now somewhat disguised by intrusive strips and slices brought under the plough with varying degrees of success." Throughout the twentieth century, the heathlands were still further fragmented, reaching almost Dalmatian spottiness and scarcity by 1978.

The British flora is, by Mexican standards, very small, in view of which you might argue that with so few treasures to start with, we must try even harder to preserve what we do have. Although Dorset heathland habitat is being extensively restored, the news overall is not so good. In the year 2000, the conservationist Peter Marren trawled through published floras for sixteen English counties and discovered to everyone's horror that, during the twentieth century, the average rate of extinction at this local level was about one species in each county each year. We do not know the precise cause, or even the exact date, when each species was lost but there is little doubt that we are the ones responsible for impoverishing our own neighborhoods. These plants are the casualties of intensive agriculture, road building, urban development, the drainage of wetlands, pollution, and nutrient enrichment. Butterflies and some birds are going the same way.

It is easy to spot an urban development that will destroy natural plant habitat, although not so easy to stop it. Nutrient enrichment is a less visible, but more insidious threat to plant diversity that extends right across the land. Nitrogen and phosphorus fertilizers applied to grasslands unleash the demon grasses that crowd out smaller, less competitive plants (chapter 6). Even in unfertilized habitats, nitrogen oxide from car exhausts and other sources is deposited from the air or dissolved out of the atmosphere by rain, adding the equivalent of forty kilograms or more of nitrogen to the vegetation in many areas (chapter 7). Thus, even in places protected from urban development, the tentacles of environmental degradation reach out to strangle plant diversity. In the Ashdown Forest in Sussex, the protected habitat of Pooh Bear and the other nursery fauna created by A. A. Milne in the 1920s, forty-seven plant species were lost between 1900 and 1996. Rivers enriched with nutrients also encourage the growth of demon plants in protected places. Many English rivers are now bordered by solid stands of stinging nettle, which luxuriates in the plentiful supply of phosphorus that they carry. How many people make the connection between the phosphate-laden washing powders they pour into their dishwasher or washing machine and the painful, impenetrable barrier that cuts them off from the enjoyment of their local river?

Every local threat to plant diversity is but a piece of the global mosaic that spells extinction. The global threat to natural habitats and wild plants and animals requires a worldwide network to keep track of it. The nerve center of this web of woe, with feelers stretching throughout the world, is improbably situated at the end of a farm track on the rural edge of the city of Cambridge, England. The office windows of the World Conservation Monitoring Centre (WCMC) overlook fields in a tranquil English landscape, but the view of global biodiversity provided on the glowing computer screens is not nearly so benign. WCMC is like the cockpit of the biosphere, whose screens are filled with images of smoke and whose dials are flickering dangerously close to red.

WCMC collates information that is entered into a database—the *Red List of Threatened Plants*—containing the names of those species known to be threatened with extinction. When published in 1997, the Red List estimated that some sixty-five hundred species are "endangered," meaning that they are in immediate danger of extinction; another eight thousand or so species are "vulnerable" and are on a slippery slope toward endangered status; and a further fourteen and one-half thousand species are classified as "rare," with populations so small or thinly scattered as to be exposed to abnormal risk. Add another four thousand species about which we know so little that all that can be said is that they hover somewhere between extinct and rare, and the needle on the dial of threatened plants shows some thirty-three thousand species to be at risk or worse. By some other estimates, the actual number of threatened species may be three times as many and could amount to nearly a quarter of all the four hundred thousand plants known.

Compiling such data is not an exact science because our knowledge of wild plant diversity is so incomplete. The number of living plant species so far described by taxonomists is thought to lie somewhere between 270,000 and 422,000, but no one knows more precisely than this how many there are. We are especially ignorant about rare species, and a good deal of the uncertainty about total species numbers is caused by rarity. It is a curious but universal fact that the majority of species in any group—be they insects, mammals, birds, fishes or plants—are naturally rare. This is so consistent a finding in every survey of living diversity that it appears to be as much a law of statistics as a law of nature. For example, half the stems in the fifty-hectare tropical forest plot on Barro Colorado Island (BCI, chapter 5) belong to a single species, but the other half are shared among over a hundred other trees and shrubs. Choose a species at random from the inventory of the BCI plot and it will usually be a rare one, but go into the forest itself and choose a tree at random and it will usually be *Trichilia tuberculata*. There are over three hundred thousand stems in the fifty-hectare plot at BCI, but some species claim fewer than twenty individuals of this number. The solution to finding such needles in haystacks at BCI was to identify and count every stem, but scale this problem up to the Island of Borneo or the Amazon basin and imagine how difficult it is to record all the rare species found there. Such are the rates of deforestation in the tropics that we can be virtually certain that many species will go extinct before they even become known to science.

Not only are rare species difficult to find and count, but they are also more vulnerable to extinction than common species because they are already dangerously near the brink, living in only a few places or in small populations. Such plants include the Wollemi pine, a tree that grows to a height of a hundred feet and that was discovered only in 1994 living in a deep canyon not seventy miles from downtown Sydney, Australia. This newly discovered rarity, uncannily resembling fossil plants thought to have been extinct for 60 million years, has a population of only forty-three adult individuals.

The inevitable undercounting of rare species is like a good-news/bad-news joke. The good news is that there are more species than you thought. The bad news is that they are rare and likely to go extinct. Because rare species are both undercounted and particularly vulnerable to extinction, the Red List undoubtedly underestimates the percentage of plant biodiversity that is really threatened.

It is easy to be pessimistic about the plight of global biodiversity, but this would be no basis for action, and action is needed if we are to save what is left. But what can be done? Ultimately it comes down to dollars. The words "ecology" and "economics" share a common etymological root in the Greek word for "home": *oikos*. Saving natural habitats, the treasure houses of biodiversity, means making economics and ecology serve each other.

The simplest idea is for the very wealthy to buy tropical habitats in less-developed countries. This is the idea behind "debt-for-nature" swaps that have been used to purchase tropical rainforest in Latin America. An organization named Conservation International (CI), based in Washington, D.C., has calculated that nearly half of all plant species and about a third of all species of mammals, birds, reptiles, and amphibians are to be found concentrated into twenty-five "hotspot" areas that add up to less than 1.5 percent of the land surface of the Earth. Purchasing these areas to protect them would cost about $23 billion, an amount comfortably less than Bill Gates's fortune. Is that it, then? Will philanthropy solve this problem? Nothing is quite so simple.

Conservation International has done remarkably well in raising several billion dollars so far and has shown that economics can serve ecology, to a degree. The key question is whether biodiversity can be sustained just by buying habitats and putting a fence around them. All the evidence is that it cannot. A great deal of the deforestation that has taken place in, for example, the rainforests of Indonesia happened illegally in " protected" areas of national park. Would CI be any more successful at protecting their property?

It seems that what must happen is that ecology must serve economics, as well as the other way around. The logic of this argument, which is well rehearsed by most organizations in the field of conservation, is that what has value for people will be protected by them. One might characterize this as the " use it or lose it" argument, though it is plainly more complicated than that. After all, the Popolucan Indians use the Mexican rainforest in a sustainable way, but it has been taken away from them. If a field of tobacco is worth more than the produce of a *milpa*, economics will sever with ecology, not serve it. The optimists believe that market forces can somehow be manipulated to make the sustainable exploitation of natural environments economically viable. Whether or not this aim is possible, the conservation of habitats and their plant biodiversity is not just about ecology, but about the relationship between plants and people. We must rediscover Eden by restoring degraded habitats. Two remarkable examples illustrate what can be done.

The renovated plumbing of the Florida Everglades ecosystem that began in the 1950s (chapter 8) brought considerable economic benefits to southern Florida, where it made possible agricultural production now worth $2 billion annually, and urban development that attracted 4.5 million people to the Miami area. Tourism also benefited and grew into an industry worth $14.5 billion to the economy of Florida in 1995. These economic benefits from drainage and flood control, however, were bought at a price to the Everglades ecosystem, which soon began to fail to deliver the services expected of it. Drainage canals diverted three or four times as much water away from southern Florida and direct into the Atlantic, as flowed by the natural route through the Everglades to the ocean. The result was that periodic droughts threatened the Everglades National Park and the drier hydrological regime caused an increase in the frequency of fires, favoring the spread of aliens such as paperbark (chapter 8). Also, the whole system no longer stored sufficient water to provide a reliable supply to the growing population of southern Florida.

Agriculture on drained land was unsustainable. Draining the once-flooded, peaty soils for agriculture exposed them to shrinkage, erosion, and oxidation—reducing soil depth in some places by as much as two meters. Run-off from agriculture also polluted drainage water with fertilizer. The added nutrients gave cattails a competitive advantage over sawgrass, replacing the main component of the vegetation in the " river of grass," with potentially widespread, though unknown, consequences for the whole ecosystem. Where the polluted water reached the ocean sea in Florida's bay, it caused algal blooms that threatened tourism, fisheries, and corals.

The good news is that in 1998, state and federal agencies committed $7.8 billion to restoring the natural hydrological balance of southern Florida, to reducing suburban growth, and to increasing wildlife habitat. Populations of the great egret have already increased in response to measures so far taken, though other wading birds such as the white ibis and the wood stork are still at low numbers. It took thirty years for the coalition of interests needed to implement these actions to come together, to agree what must be done, and to vote the funds for the largest ecological restoration project so far undertaken anywhere. Even more money will be needed before all the project's aims are met, perhaps in another thirty years' time.

The idea of restoring damaged ecosystems to haul biodiversity back from the brink now enjoys widespread support among conservationists, but it was not always so. Restoration seemed particularly unpromising for saving tropical forest diversity because it was thought that deforestation irreversibly depleted the nutrients on which plant growth depends. In the Amazon, for example, where forest grows on very nutrient-poor soils, trees survive by very tight recycling of nutrients. Virtually all the nutrients needed for plant growth in such an ecosystem are locked-up in the plants themselves. Destroy the vegetation and you lose the nutrients too. Perhaps it is for this reason that, back in 1985, when Dan Janzen , he of the Janzen-Connell hypothesis (chapter 5), stood up at a meeting in Washington, D.C. on saving tropical forests and proposed regrowing them, his suggestion was met with considerable scepticism. Not only was tropical forest restoration thought to be ecologically impractical, but some conservationists were worried that the very idea that tropical deforestation might be reversible would undermine their fundraising efforts, which were based on the very opposite assumption.

Ever the iconoclast, Janzen pointed out at the meeting that the assumption that deforestation is irreversible had never been tested. For one thing, not all tropical forests grow on poor soils, particularly in the volcanic regions of Mexico and Central America where Janzen worked. This meeting was a turning point for Janzen and led indirectly to the initiation of a project at Guanacaste in Costa Rica that has become a beacon of tropical forest restoration. How had Janzen arrived at this juncture from a highly successful career in basic research? I visited him at his lab at the University of Pennsylvania in Philadelphia in 1983 and found a man totally driven by his subject. He had already published some 250 scientific articles, twice the number most scientists achieve in their entire career, and he was engaged in a project to describe the moths of Costa Rica, a mammoth task that continues to this day.

Dan explained that he had decided to divert some of his time toward saving the tropical forests of Costa Rica where he conducted his research. As he later told the science writer William Allen:

> The problem is, I could sit here and stack up an enormous amount of very pretty research over a period of the next thirty years or so that I have left in me, and that's fine. And then the next generation of people will turn around and look at this forest and it will all be cow pasture. So, what's better? Spending my time stacking up thirty years of research, or only doing half as much research and ending up saving the place, and leaving it for other people to do vastly more research? Because what I'll get done in my lifetime is still tiny compared to what the potential of this place is.

Around the same time, Janzen used an article he titled " The Future of Tropical Ecology," published in the *Annual Review of Ecology and Systematics*, an academic journal not normally given to campaigning, to encourage other tropical ecologists that they should do likewise: " Set aside your random research and devote your life to activities that will bring the world to understand that tropical nature is an integral part of human life. If our generation does not do it, it won't be there for the next."

When Janzen issued this call to arms, he had been working in

Costa Rica for many years from a base at Santa Rosa National Park in Guanacaste province in the north of the country, near its border with Nicaragua. The natural vegetation of the lowlands in this region is tropical dry forest, not as tall in stature nor as rich in species as rainforest, but rich in plants and animals nonetheless. Rainfall there is modest and seasonal, and about half the tree species are leafless in the dry season. The reduced stature of the forest accommodates only two stories of trees—a sparse upper layer of canopy species between twenty and thirty meters tall with stout trunks and spreading crowns, and a lower layer of understory trees between ten and twenty meters tall. Beneath these is a shrub layer of sometimes dense, thorny scrub that can be difficult to penetrate. Among the canopy species grows the chichle tree whose latex is the original source of chewing gum. Some old chichle trees in Santa Rosa still bear the criss-cross scars left in their bark by the *chicleros* (latex collectors) who once tapped the trunks for gum. A more common land use in the last fifty years has been cattle ranching and it was this encroaching threat to the forest that spurred Dan to action.

Ranchers cleared the forest around Santa Rosa and introduced African and Australian grasses on which to pasture their cattle. The artificial savannah grassland produced by cutting, burning, and grazing the forest became completely dominated by an African grass called *Jaraguá* that is native to the plains of the Serengeti. This demon grass has an infernal weapon in its armory: it is both highly flammable and fire-tolerant. It explodes into flame when torched during the dry season and then regrows with redoubled vigor. Woody plants, particularly young ones, are eliminated in the inferno. Natural fires are not frequent in the dry forest, but ranchers light them to stimulate the growth of *Jaraguá* grass and to kill woody seedlings, thus preventing the forest from returning.

Working in Santa Rosa, Dan Janzen could see that when fire was prevented, trees did begin to return to the *Jaraguá* pastures. There is a natural tendency for vegetation to transform itself, as herbaceous and low-growing plants are replaced by woody, taller species. Given enough time, this process, known as "succession," can eventually restore forest; once the plants have become established, the forest animals follow. However, succession can only get started if there is intact forest somewhere within reach to provide a source of seeds. Seeds can travel hundreds of meters, aided by wind or by fruit-eating birds and mammals, but the farther they must travel, the fewer the species that can actually reach a restoration site and the more likely it is that the site will become dominated by just a few very mobile tree species. The only way around this problem is to give nature a helping hand by deliberately planting some species or by using domestic animals to aid seed dispersal. When Janzen started to raise money to buy land for restoration in Guanacaste, his first flyer showed a newly germinated seedling of the eponymous Guanacaste tree sprouting from a cowpat.

A tree sprouting from a cowpat represented both the means and the objective of restoration: use cows to turn pasture into forest. This sounds contradictory, but it worked so long as cow densities were kept low and fire was kept out. In 1988 Janzen wrote that his recipe for ecological restoration in the tropics was to work with what was already to hand, including ranchers and their cattle:

> Choose an appropriate site, obtain it, and hire some of the former users as live-in managers. Sort through the habitat remnants to see which can recover. Stop the biotic and physical challenges to those remnants. The challenge is to turn the farmer's skill at biomanipulation to work for the conservation of biodiversity.

In Guanacaste, the chief biotic threat was overgrazing and the physical threat was fire. Both of these arose from the existing land use, but it was not enough just to purchase farmers' land. Conservation of the forest had to be embedded in the local economy or there would be a potential social threat. Social acceptance was promoted by hiring ex-farmers and other locals as firefighters, park guards, guides, or for work in other roles including research and education. Twelve years later, Santa Rosa National Park served as the nucleus of a much larger zone covering 120,000 hectares of regenerating dry forest, designated the Guanacaste Conservation Area (in Spanish: Area de Conservacion Guanacaste or ACG). Janzen wrote: " Restoring complex tropical wildlands is primarily a social endeavour; the technical issues are far less challenging."

Success of the ACG depended on a holistic approach to conservation that took account of its social as well as its ecological dimensions. The social endeavor that created the ACG is chronicled in detail by William Allen in his book *Green Phoenix*. He makes it clear that a key ingredient in the success of the project was a large injection of cash, sufficient to turn ecology in Guanacaste into a self-sustaining business that in 1999 had an operating budget of $1.6 million. The ACG generates income from eco-tourism, from the provision of research services, and from licences with pharmaceutical companies looking for leads to new drugs from forest plants and animals, but the budget is underpinned by a $24.5 million endowment. It will take centuries for ecological succession to fully restore mature forest in the ACG and it is essential that the project is economically viable, or it will fail.

What are the lessons of the ACG? There are at least three. First is Janzen's dictum, " hire some of the former users as live-in managers," or more generally, find sustainable uses for natural or restored habitats that will give local people an economic incentive to protect biodiversity. A sustainable use is one that generates some economic return without diminishing the capacity of the ecosystem to produce similar returns in the future. These uses will be different from place to place, but the philosophy of " use it or lose it" is being successfully adopted in many places, from Europe, where farmers can obtain subsidies for ecological management of their land, to the Himalayas, where reforestation protects watersheds from erosion and forest products provide new sources of income for local people. Designing sustainable uses that protect biodiversity is not easy and can only be done by understanding how diversity is maintained and the nature of threats it faces. For example, species-rich grassland must be grazed, but it should not be fertilized (chapter 6).

The second lesson is that the demons in Eden that stifle plant diversity, such as *Jaraguá* grass, are often there by invitation. They can and must be removed. In the ACG, extermination was achieved relatively easily by controlling fire; in Florida or Hawaii, however, where there are whole armies of demons, the task is much more difficult. Even when the hydrology of the Everglades ecosystem has been restored, the war with Florida's demons will not have been won. To conquer them, we must understand what turns plants into demons in the first place.

The third lesson is a message of real hope: what humans have damaged, nature can heal, given help and time. Once fire and *Jaraguá* had been controlled, dry forest species began rapidly to reinvade the old pastures of ACG. Succession will restore other habitats wherever there are fragments to act as seed sources. But small habitat fragments have an impoverished flora and fauna, so we must save as much natural habitat as possible and use smaller fragments to rebuild larger areas that can support viable populations of more species. Succession is a phased process and vegetation in its earliest stages does not support the diversity of species to be found in mature habitat. The forest now covering the former pastures of the ACG is dominated by trees with wind-dispersed seeds, because these were the first arrivals. Yet there are many more tree species waiting in the wings and these are already creeping from their refuges into the new forest. With time, the forest will diversify because diversity is the natural state of things. Some species may have been permanently lost, but with the demons under control, diversity will return.

Guanacaste, where people in league with nature do battle with demons in defense of biodiversity, is a fitting departure point for a final reprise of the theme of this book. At the start of this journey we discovered the extraordinary diversity of plants in Kew Gardens, which, by chance, has historical links with Charles Darwin. Darwin's ideas on evolution by natural selection are fundamental to our understanding of why there are so many kinds of plants. He well understood the power of population increase and the tendency inherent in every population to multiply geometrically when circumstances are favorable. Thus natural selection can catapult new forms from obscurity to dominance in just a handful of generations, producing evolutionary change. These are the plants I have termed "Darwinian demons." All species must go through a demon phase in their evolutionary history, but this fact produces a paradox: if every population has demonic potential, how does diversity arise? Why don't new species simply wipe out old ones?

Darwin puzzled over this paradox himself and when the answer finally dawned on him, it left such an impression that years later he still remembered "the very spot in the road, whilst in my carriage, when to my joy the solution occurred to me." The solution is that new species find refuge from old demons by invading new ecological niches. The flora of the Canary Islands shows us that when new habitats are colonized, evolution spills forth a cornucopia of new species.

Even where they are at their most successful, demons, like the superheroes of modern DC comics and ancient B.C. Greece, eventually reveal their fatal weaknesses. On Mount Shimagare in Japan, where a handful of species dominate the sub-alpine zone, we met the demon bamboo *Sasa* that is so powerful that it can check the growth of trees until the bamboo itself flowers and then dies. For sub-alpine fir trees too, sex is their undoing and limits growth. The general point is that every organism is subject to trade-offs that prevent it from excelling in every way in every environment. Trade-offs lead to specialization, and that is the key to diversity. In Panama, specialization of various kinds permits very large numbers of plant species to coexist in tropical forest. Tropical trees and their numerous natural enemies have mutually specialized abilities of defense and attack. This helps maintain forest diversity by favoring rare species. In meadows, plants specialize along subtle gradients of soil moisture. Here, and in other grasslands, mowing and grazing animals are also crucial to diversity because they limit the growth of potentially dominant plants, especially grasses.

Wherever it occurs, the coexistence of large numbers of species depends on a delicate balance of trade-offs to prohibit the destructive power of demons. This balance is easily disrupted, with dire consequences. A spreading and insidious global threat is nutrient pollution of natural habitats, especially by atmospheric nitrogen deposition. This has already damaged biodiversity in Europe, North America, and elsewhere and there is almost certainly much more damage from the same source in the pipeline (and the tailpipe).

Invading species transported to new areas without their natural enemies are another threat to natural habitats, especially in the New World. The ecological disasters that have been caused by invasive plants raise the fear that genetic engineering might produce new breeds of demon. The threat is real, but arises from the new traits that genetically modified plants might be engineered with, rather than from any inherent risk in the technology itself.

Two hundred million years of evolution have given us more species of flowering plants than the world has ever known. Must we fall from this pinnacle of diversity into a pit of destruction? Will the Darwinian demons, held in check for so long by nature's laws, conquer at last through human folly and indifference? They may, and in places they already have; yet it is not too late for action. What is needed is a determination to leave our children a living inheritance that has not been diminished either by our actions, or by our failure to act.

Scientific Names of Plants Mentioned in the Text

I use the common names for plants mentioned in the text wherever possible. In some places, however, I give both the common and scientific names because the latter indicate an important relationship with other plants mentioned, as in the case of hybrids or species belonging to the same genus. In the list below, the abbreviation "spp." denotes that more than one species is covered by the common name given. The abbreviation "sp." denotes one species of unknown identity, "subsp." means subspecies, and "var." refers to a variety.

COMMON NAME	SCIENTIFIC NAME
Alaskan cedar	*Chamaecyparis nootkatensis*
Alligator weed	*Alternanthera philoxeroides*
American beech	*Fagus grandifolia*
American cord grass	*Spartina alternifolia*
Australian pine	*Casuarina* spp.
Autumn gentian	*Gentianella amarella*
Balsam	*Impatiens* spp.
Balsam fir	*Abies balsamea*
Beach jacquemontia	*Jacquemontia reclinata*
Beach star	*Remirea maritima*
Bean	*Phaseolus* spp.
Bee orchid	*Orchis apifera*
Birds-foot trefoil	*Lotus corniculatus*
Brazilian pepper	*Schinus terebinthifolius*
Briar rose	*Rosa* sp.
Bulbous buttercup	*Ranunculus bulbosus*
Buttercup	*Ranunculus* spp.
Cabbage	*Brassica oleracea*
Canary olive	*Olea europaea* sp. *Cerasiformi*
Canary pine	*Pinus canariensis*
Canola (oilseed rape)	*Brassica napus*
Carnation	*Dianthus* spp.
Cattail	*Typha* spp.
Century plant	*Agave americana*
Chicle	*Manilkara zapota*

COMMON NAME	SCIENTIFIC NAME
Chilli pepper	*Capsicum frutescens*
Chinese water spinach	*Ipomoea repens*
Cocksfoot grass	*Dactylis glomerata*
Cocoa	*Theobroma cacao*
Coffee	*Coffea arabica*
Cogon grass	*Imperata cylindrica*
Common spotted orchid	*Dactylorhiza fuchsii*
Common sunflower	*Helianthus annuus*
Cord grass	*Spartina* spp.
Cowslip	*Primula veris*
Creeping buttercup	*Ranunculus repens*
Cycad	*Zamia pumila*
Diffuse knapweed	*Centaurea diffusa*
Dragon tree	*Dracaena draco*
Dropwort	*Filipendula vulgaris*
Dwarf bamboo	*Sasa nipponica*
Dwarf centaury	*Centaurium erythrea*
Dwarf cornel	*Cornus canadensis*
Early purple orchid	*Orchis mascula*
English cord grass	*Spartina anglica*
Fairy flax	*Linum catharticum*
Field scabious	*Knautia arvensis*
Florida arrowroot	*Zamia pumila*
Florida rosemary	*Ceratiola ericoides*
Florida slash pine	*Pinus elliottii* var. *densa*
Gentians	*Gentiana* spp.
Geranium	*Pelargonium* spp.
Giant Amazon water lily	*Victoria amazonica*
Giant bamboo	*Phyllostachys bambusoides*
Gorse	*Ulex europaeus*
Greater knapweed	*Centaurea scabiosa*
Ground cherry	*Physalis angustifolia*
Guava	*Psidium guajava*
Hawkbit	*Leontodon* spp.
Himalayan balsam	*Impatiens glandulifera*
Hosta	*Hosta* spp.
Houseleek	*Aeonium* spp.
Hydrilla	*Hydrilla verticillata*
Jade plant	*Crassula argentea*
Japanese honeysuckle	*Lonicera japonica*
Japanese knotweed	*Reynoutria japonica*
Jaraguá grass	*Hyparrhenia rufa*
Kentucky bluegrass	*Poa pratensis*
Klamath weed	*Hypericum perforatum*

COMMON NAME	SCIENTIFIC NAME
Knapweed	*Centaurea* spp.
Kudzu	*Pueraria lobata*
Lacandonia	*Lacandonia schismatica*
Ladies' bedstraw	*Galium verum*
Lantana	*Lantana camara*
Laurel	*Laurus* spp.
Laurel fig	*Ficus microcarpa*
Lesser burnet saxifrage	*Pimpinella saxifraga*
Lesser knapweed	*Centuarea nigra*
Leyland cyprus	*Cupressocyparis leylandii*
Living stones	*Lithops* spp.
Maize (corn)	*Zea mays*
Maple	*Acer* spp.
Medic	*Medicago lupulina*
Monterey cyprus	*Cupressus macrocarpa*
Oak	*Quercus* spp.
Old World climbing fern	*Ligodium microphyllum*
Orange balsam	*Impatiens capensis*
Oxeye daisy	*Leucanthemum vulgare*
Paperbark	*Melaleuca quinquenervia*
Pelican flower	*Aristolochia grandiflora*
Pirri-pirri bur	*Acaena novae-zelandiae*
Plantain	*Plantago* spp.
Pond apple	*Annona glabra*
Prairie sunflower	*Helianthus petiolaris*
Primrose	*Primula* spp.
Purple loosestrife	*Lythrum salicaria*
Red maple	*Acer rubrum*
Red pine	*Pinus Resinosa*
Red spruce	*Picea rubens*
Restharrow	*Ononis repens*
Rhododendron	*Rhododendron ponticum*
Rockland ruellia	*Ruellia succulenta*
Sawgrass	*Cladium jamaicense*
Sedge	*Carex* spp.
Sheep's fescue grass	*Festuca ovina*
Shoebutton ardisia	*Ardisia elliptica*
Smooth-stalked meadow grass	*Poa pratensis*
Snake's-head fritillary	*Fritillaria meleagris*
Sorrel	*Oxalis* spp.
Sow thistle	*Sonchus* spp.
Squash	*Cucurbita* spp.
St. John's wort	*Hypericum perforatum*
Tall buttercup	*Ranunculus acris*

COMMON NAME	SCIENTIFIC NAME
Titan arum	*Amorphophallus titanum*
Tor grass	*Brachypodium pinnatum*
Violet	*Viola* spp.
Water hyacinth	*Eichhornia crassipes*
Water lettuce	*Pistia stratiotes*
White clover	*Trifolium repens*
White Marguerite	*Argyranthemum frutescens*
Wild carrot	*Daucus carota*
Wild parsnip	*Pastinaca sativa*
Wild thyme	*Thymus serpillifolium*
Yarrow	*Achillea millefolium*
Yellow-wort	*Blackstonia perfoliata*

Sources and Further Reading

See generally www.demonsineden.com

PREFACE

Notes

ix . . . 40 percent of the plant growth . . . : S. Pimm, *The World according to Pimm: A Scientist Audits the Earth* (New York: McGraw Hill, 2001).

ix Today there are approximately four hundred thousand flowering plant species . . . : R. Govaerts, "How Many Species of Seed Plants Are There?" *Taxon* 50 (2001): 1085–90.

ix . . . a quarter of all plant species are presently at risk . . . : B. Groombridge and M. D. Jenkins, *Global Biodiversity: Earth's Living Resources in the 21st Century* (Cambridge: World Conservation Press, 2000).

ix . . . we are also kin to our pet goldfish and dogs, to the tomatoes . . . : Richard Dawkins gives an excellent account of the evolutionary unity of all life in his book *The Ancestor's Tale* (London: Weidenfeld and Nicholson, 2004).

CHAPTER 1: AN EVOLVING EDEN

General Sources

There is no substitute for an actual visit to Kew Gardens, but Kew's website has plenty to offer: www.rbgkew.org.uk.

Erasmus Darwin is a fascinating character and too little known. I recommend the biography by D. King-Hele, *Erasmus Darwin: A Life of Unequalled Achievement* (London: De La Mare Publishers, 1999). His botanical and evolutionary poetry has been republished in a modern edition titled *Cosmologia*, edited by Stuart Harris and available from the Erasmus Darwin Society's website at www.erasmusdarwin.org.

There is no shortage of biographies of Charles Darwin. I have mainly used volume 2 of Janet Browne's two-volume biography *Charles Darwin: The Power of Place* (London: Jonathan Cape, 2002).

Notes

1 . . . there "[s]its enthroned in vegetable pride . . .": *Essential Writings of Erasmus Darwin*, ed. D. King-Hele (London: McGibbon and Kee, 1968).

4 Sir Ghillean Prance, . . . discovered how the giant lily is pollinated . . . : C. Langmead, *A Passion for Plants: The Life and Vision of Ghillean Prance* (Oxford: Lion Publishing, 1995).

7 "co-circum-wanderer and fellow labourer": Letter of 23 Feb. 1844, in *Charles Darwin's Letters: A Selection*, ed. F. Burkhardt (Cambridge: Cambridge University Press, 1996).

7 Darwin's grandfather Erasmus had been publicly ridiculed . . . : King-Hele, *Erasmus Darwin.*

8 "the generous assistance which I have received from very many naturalists . . .": Introduction to C. Darwin, *The Origin of Species by Means of Natural Selection*, 1st ed. (London: John Murray, 1859; reprinted, New York: Penguin Books, 1974).

8 . . . Erasmus had propounded it in verse in his 1803 poem *The Temple of Nature:* King-Hele, *Erasmus Darwin.*

8 "I think I have found out (here's presumption!) the simple way by which species become exquisitely adapted to various ends": Letter of 11 Jan. 1844, in *Charles Darwin's Letters.*

8 In wild radish, for example, . . . : T. N. Lee and A. A. Snow, "Pollinator Preferences and the Persistence of Crop Genes in Wild Radish Populations (*Raphanus raphanistrum*, Brassicaceae), *American Journal of Botany* 85 (1998): 333–39.

8 "A struggle for existence inevitably follows . . .": Darwin, *The Origin*, ch. 3.

9 "Each pregnant Oak ten thousand acorns forms . . .": Canto IV, *The Temple of Nature*, in *Essential Writings of Erasmus Darwin.*

9 "I use the term Struggle for Existence . . .": Darwin, *The Origin*, ch. 3.

9 "Linnaeus has calculated . . .": ibid.

9 "Cases could be given of introduced plants . . .": ibid.

10 . . . Leyland cypress . . . is a hybrid . . . : M. Campbell-Culver, *The Origin of Plants* (London: Hodder, 2001).

11 "the subject of *leylandii* has dominated my postbag . . .": Michael Meacher, quoted in the *Guardian*, 13 Nov. 1999.

CHAPTER 2: THE TREE OF TREES

General Sources

The tree of trees is merely (I do not really mean merely, of course) the green branch of the tree of life. Our knowledge of the structure of this larger tree is growing so rapidly that no printed work can hope to be fully up to date, even on the day it is published. Even websites struggle to keep up, but the one to keep an eye on is the Tree of Life website at www.tol.web.

For the latest taxonomic treatment of the flowering plants based on phylogenetics, visit the Angisoperm Phylogeny Group website: www.apg.org.

Notes

14 "As buds give rise . . .": C. Darwin, *The Origin of Species by Means of Natural Selection*, 1st ed. (London: John Murray, 1859); reprinted, New York: Penguin Books, 1974), ch. 4.

17 "nothing could equal the gross prurience of Linnaeus' mind": W. C. Stearn, in W. Blunt, *The Compleat Naturalist* (London: Collins, 1971), 245.

17 "What Beaux and Beauties crowd the gaudy groves . . .": *Essential Writings of Erasmus Darwin*, ed. D. King-Hele (London: McGibbon and Kee, 1968).

19 . . . a collaboration involving forty-one other scientists": M. W. Chase et al., "Phylogenetics of Seed Plants: An Analysis of Nucleotide-Sequences from the Plastid Gene *rbc*L," *Annals of the Missouri Botanical Garden* 80 (1993): 528–80.

20 "persisted in the notion . . .": M. W. Chase and V. A. Albert, "A Perspective on the Contribution of Plastid *rbc*L DNA Sequences to Angiosperm Phylogenetics," in *Molecular Systematics of Plants*, vol. 2, ed. P. S. Soltis, D. E. Soltis, and J. J. Doyle (London: Chapman and Hall, 1998).

24 . . . a three-gene tree . . . : P. S. Soltis, D. E. Soltis, and M. W. Chase, "Angiosperm Phylogeny Inferred from Multiple Genes as a Tool for Comparative Biology," *Nature* (1999): 402, 402–4.

24 . . . William Bateson at the University of Cambridge . . . : R. M. Henig, *A Monk and Two Peas* (London: Weidenfeld and Nicholson, 2000).

25 A tiny, three-millimeter-long water lily flower . . . : E. M. Friis, K. R. Pedersen, and P. R. Crane, "Fossil Evidence of Water Lilies (*Nymphaeales*) in the Early Cretaceous," *Nature* (2001): 410, 357–60.

CHAPTER 3: SUCCULENT ISLES

General Sources

The definitive, illustrated flora of the Canary Islands is D. Bramwell and Z. Bramwell, *Wild Flowers of the Canary Islands* (Madrid: Rueda, 2001) (also available in Spanish.). A rich scientific account of the natural history of the Canaries written mainly by scientists from the region is J. M. Fernández-Palacios and J. L. M. Esquivel, eds., *Naturaleza de las Islas Canarias* (Santa Cruz de Tenerife: Turquesa, 2001).

Further details and evidence for the argument presented in this chapter that competition inhibited multiple colonization in the Canaries is in J. Silvertown, "The Ghost of Competition Past in the Phylogeny of Island Endemic Plants," *Journal of Ecology* 92 (2004): 168–73.

Notes

30 An analysis of DNA from populations of the olive . . . : J. Hess, J. W. Kadereit, and P. Vargas, "The Colonization History of *Olea europaea* L. in Macaronesia Based on Internal Transcribed Spacer 1 (ITS-1) Sequences, Randomly Amplified Polymorphic DNAs (RAPD), and Intersimple Sequence Repeats (ISSR)," *Molecular Ecology* 9 (2001): 857–68.

32 . . . Canarian endemics related to the sow thistles . . . : S. C. Kim et al., "A Common Origin for Woody Sonchus and 5 Related Genera in the Macaronesian Islands: Molecular Evidence for extensive radiation," *Proceedings of the National Academy of Sciences of the USA* 93 (1996): 7743–48.

35 . . . the twenty-three *Argyranthemum* species descend from a single colonization . . . : J. Francisco-Ortega, R. K. Jansen, and A. Santos-Guerra, "Chloroplast DNA Evidence of Colonization, Adaptive Radiation, and Hybridization in the Evolution of the Macaronesian Flora," *Proceedings of the National Academy of Sciences of the USA* 93 (1996): 4085–90.

37 "the tendency in organic beings descended from the same stock to diverge in character as they become modified": *Life of Charles Darwin*, ed. F. Darwin (London: John Murray, 1902).

38 "I can remember the very spot in the road . . .": ibid.

42 Sixty-nine Canary Island endemics . . . : C. Hobohm, "Plant Species Diversity and Endemism on Islands and Archipelagos, with Special Reference to the Macaronesian Islands," *Flora* 195 (2000): 9–24.

CHAPTER 4: DEMON MOUNTAIN

General Sources

Many of the topics covered in chapter 4 are dealt with in more detail in J. Silvertown and D. Charlesworth, *Introduction to Plant Population Biology*, 4th ed. (Oxford: Blackwell Science, 2001). Chapter 10 of that book deals with the evolution of life history in plants.

Notes

44 . . . coniferous trees in the Northern Hemiphere synchronize their seed production . . . : W. D. Koenig and J. M. H. Knops, "Scale of Mast-Seeding and Tree-Ring growth," *Nature* 396 (1998): 225–26.

44 Parts of the same bamboo clone flower and die in synchrony . . . : D. H. Janzen, "Why Bamboos Wait So Long to Flower," *Annual Review of Ecology and Systematics* 7 (1976): 347–91.

50 Takashi Kohyama . . . studied fir waves on Mount Shimagare . . . : T. Kohyama and N. Fujita, "Studies on the *Abies* Population of Mt. Shimagare. I. Survivorship Curve," *Botanical Magazine, Tokyo* 94 (1981): 55–68; T. Kohyama, "Studies on the *Abies* Population of Mt. Shimagare. II. Reproductive and Life History Traits," *Botanical Magazine, Tokyo* 95 (1982): 167–81.

53 . . . a computer model created by . . . Sato and Yoh Iwasa . . . : K. Sato and Y. Iwasa, "Modelling of Wave Regeneration in Sub-Alpine *Abies* Forests: Population Dynamics with Spatial Structure," *Ecology* 74 (1993): 1538–50.

55 The two species of *Abies* coexist . . . : T. Kohyama, "Regeneration and Coexistence of Two *Abies* Species Dominating Subalpine Forests in Central Japan," *Oecologia* 62 (1984): 156–61.

56 It was at Whiteface Mountain that we discovered . . . : J. Silvertown and M. E. Dodd, "The Demographic Cost of Reproduction and Its Consequences in Balsam Fir (*Abies balsamea*)," *American Naturalist* 154 (1999): 321–32; J. Silvertown and M. E. Dodd, "Evolution of Life History in Balsam Fir (*Abies balsamea*) in Sub-Alpine Forests," *Proceedings of the Royal Society of London* 266 (1999): 729–33.

CHAPTER 5: THE PANAMA PARADOX

General Sources

A comprehensive account of the ecology of Barro Colorado Island is E. G. J. Leigh, A. S. Rand, and D. M. Windsor, eds., *The Ecology of a Tropical Forest* (Washington, D.C.: Smithsonian Institution Press, 1982). Steve Hubbell gives a summary of research in the fifty-hectare plot at BCI in S. P. Hubbell, "Two Decades of Research on the BCI Forest Dynamics Plot," in *Tropical Forest Diversity and Dynamism*, ed. E. Losos and E. G. Leigh, 8–30 (Chicago: University of Chicago Press, 2004).

Notes

61 . . . Darwin's finches in the Galapagos Islands . . . : J. Weiner, *The Beak of the Finch* (London: Vintage, 1996).

62 "And NUH is the letter I use to spell Nutches . . .": Dr. Seuss, *On Beyond Zebra* (New York: Random House, 1955).

62 In 1970 Dan Janzen . . . : D. H. Janzen, "Herbivores and the Number of Tree Species in Tropical Forest," *American Naturalist* 104 (1970): 501–28.

62 . . . and Jo Connell . . . : J. H. Connell, On the Role of Natural Enemies in Preventing Competitive Exclusion in Some Marine Animals and in Rain Forests," in *Dynamics of Populations*, ed. P. J. den Boer and G. R. Gradwell, 298–310 (Wageningen, Netherlands: Centre for Agricultural Publishing and Documentation, 1971).

63 A significant proportion of our medicines . . . : A. Beattie and P. R. Ehrlich, *Wild Solutions* (New Haven: Yale University Press, 2001).

64 "When we finished the first census I was utterly surprised at the vast number of stems": Steve Hubbell, transcript of a television interview, program S328/1, *Tropical Forest: The Conundrum of Coexistence.* Open University copyright. Filmed in 1995.

65 In 1986 Steve Hubbell and Robin Foster . . . : S. P. Hubbell and R. B. Foster, "Biology, Chance, and History and the Structure of Tropical Rain Forest Tree Communities," in *Community Ecology*, ed. J. Diamond and T. J. Case, 314–29 (New York: Harper Row, 1986); S. P. Hubbell, *The Unified Neutral Theory of Biodiversity and Biogeography* (Princeton: Princeton University Press, 2001).

66 . . . Goren Ågren and Toby Fagerström . . . : G. I. Ågren and T. Fagerström, "Limiting Dissimilarity in Plants: Randomness Prevents Exclusion of Species with Similar Competitive Abilities," *Oikos* 43 (1984): 369–75.

67 Sixty-five species have disappeared . . . : W. D. Robinson, "Long-term Changes in the Avifauna of Barro Colorado Island, Panama: A Tropical Forest Isolate," *Conservation Biology* 13 (1999): 85–97.

68 . . . in 1977 Bill Hamilton and Bob May . . . : W. D. Hamilton and R. M. May, "Dispersal in Stable Habitats," *Nature* 269 (1977): 578–81.

69 . . . the effect of poaching of the mammals . . . : S. J. Wright et al., "Poachers Alter Mammal Abundance, Seed Dispersal, and Seed Predation in a Neotropical Forest," *Conservation Biology* 14 (2000): 227–39.

69 . . . Darwin wrote a whole book about them: C. Darwin, *The Movements and Habits of Climbing Plants* (London: Murray, 1905).

70 . . . the same fungal clone for 23 million years: I. H. Chapela et al., "Evolutionary History of the Symbiosis between Fungus-growing Ants and Their Fungi," *Science* 266 (1994): 1691–94.

71 The probable cause is global warming: R. Condit, S. P. Hubbell, and R. B. Foster, "Changes in Tree Species Abundance in a Neotropical Forest: Impact of Climate-Change," *Journal of Tropical Ecology* 12 (1996): 231–56.

71 A study of these colonies . . . : D. H. Janzen, ed., *Costa Rican Natural History* (Chicago: Chicago University Press, 1983), 752.

73 A very important clue has come . . . : K. E. Harms et al., "Pervasive Density-Dependent Recruitment Enhances Seedling Diversity in a Tropical Forest," *Nature* 404 (2000): 493–95.

73 As the data from repeat censuses became available . . . : J. A. Ahumada et al., "Long-term Tree Survival in a Neotropical Forest," in *Tropical Forest Diversity and Dynamism*, ed. E. Losos and E. G. Leigh, 408–32 (Chicago: Chicago University Press, 2004).

73 "plants that we thought were young . . .": Hubbell, transcript of television interview.

74 . . . a palm called *Scheelea zonensis* . . . : S. J. Wright, "The Dispersion of Eggs by a Bruchid Beetle among *Scheelea* Palm Seeds and the Effect of Distance to the Parent Palm," *Ecology* 64 (1983): 1016–21.

74 . . . a canker disease affecting the large tree *Ocotea whitei* . . . : G. S. Gilbert et al., "Effects of Seedling Size, El Nino Drought, Seedling Density, and Distance to Nearest Conspecific Adult on 6-year Survival of *Ocotea whitei* Seedlings in Panama," *Oecologia* 127 (2001): 509–16.

CHAPTER 6: NIX NITCH

General Sources

I have summarized recent evidence for niche separation in plants in J. Silvertown, "Plant Coexistence and the Niche," *Trends in Ecology and Evolution* 19(11) (2004): 605–11. A broader view of ecological niches is J. M. Chase and M. A. Leibold, *Ecological Niches* (Chicago: University of Chicago Press, 2003). The theory of coexistence is reviewed in P. Chesson, "Mechanisms of Maintenance of Species Diversity," *Annual Review of Ecology and Systematics* 31 (2000): 343–66.

Notes

76 "He rises and begins to round . . .": *The Lark Ascending*, in *The Poems of George Meredith*, ed. P. B. Bartlett, 288 (New Haven: Yale University Press, 1978).

81 Peter Grubb suggested . . . : P. J. Grubb, "The Maintenance of Species Richness in Plant Communities: The Importance of the Regeneration Niche," *Biological Reviews* 52 (1997): 107–45.

81 "As species of the same genus have usually . . .": C. Darwin, *The Origin of Species by Means of Natural Selection*, 1st ed. (London: John Murray, 1859), ch. 3.

81 . . . I chose to study three pairs of species . . . : J. Silvertown and F. R. Wilkin, "An Experimental Test of the Role of Micro-Spatial Heterogeneity in the Co-existence of Congeneric Plants," *Biological Journal of the Linnean Society* 19 (1983): 1–8.

82 . . . eventually led me to the conclusion . . . : J. Silvertown and R. Law, "Do Plants Need Niches? *Trends in Ecology and Evolution* 2 (1987): 24–26.

87 . . . Heinz Ellenberg had constructed . . . : H. Ellenberg, "Physiologisches und ökologisches verhalten derselben Pflanzenarten," *Berichter der Deutschen Botanischen Gesellschaft* 65 (1953): 350–61.

89 Publication in *Nature* . . . : J. Silvertown et al., "Hydrologically-Defined Niches Reveal a Basis for Species-Richness in Plant Communities," *Nature* 400 (1999): 61–63.

89 . . . between different oak species in Florida . . . : J. Cavender-Bares et al., "Phylogenetic Overdispersion in Floridian Oak Communities," *American Naturalist* 163 (2004): 823–43.

CHAPTER 7: LIEBIG'S REVENGE

General Sources

A scholarly biography of Justus von Liebig in English is W. H. Brock, *Justus von Liebig: The Chemical Gatekeeper* (Cambridge: Cambridge University Press, 1997). The question of where all the extra nitrogen that industrial society uses comes from, which I have not mentioned here, is dealt with in G. J. Leigh, *The World's Greatest Fix: A History of Nitrogen and Agriculture* (Oxford: Oxford University Press, 2004).

Notes

91 "J. B. LAWE'S PATENT MANURES . . .": A. E. Johnston, "Liebig and the Rothamsted Experiments," in *Symposium 150 Jahre Agrikulturchemie*, ed. G. K. Judel and M. Winnewisser, 37–64 (Giessen: Justus Liebig-Gesellschaft zu Giessen, 1991).

92 "[T]he plots had each so distinctive a character . . .": quoted in E. D. Williams, *Botanical Composition of the Park Grass Plots at Rothamsted 1856–1976* (Harpenden: Rothamsted Experimental Station, 1978).

92 Botanical sampling to record the differences . . . : J. B. Lawes and J. H. Gilbert, "Agricultural, Botanical and Chemical Results of Experiments on the Mixed Herbage of Permanent Meadow, Conducted for More Than Twenty Years in Succession on the Same Land. Part I: The Agricultural Results," *Philosophical Transactions of the Royal Society (A & B)* 171 (1880): 289–415; "Part II: The Botanical Results," *Philosophical Transactions of the Royal Society (A & B)* 173 (1880): 1181–1413.

92 "Some plots were much more difficult . . .": E. Grey, *Rothamsted Experimental Station: Reminiscences, Tales and Anecdotes of the Laboratories, Staff and Experimental Fields 1872–1922*. Privately published by E. Grey, Laboratory Cottage, Harpenden, Herts.

93 . . . "Johnny" Johnston . . . has speculated . . . : Johnston, "Liebig and the Rothamsted Experiments."

93 Liebig was the foremost authority on agricultural chemistry . . . : Brock, *Justus von Liebig.*

93 . . . a report on agricultural chemistry . . . : J. von Liebig, *Organic Chemistry in Its Applications to Agriculture and Physiology*, trans. L. Playfair (London: n.p., 1840).

93 "It must be admitted as a principle of agriculture . . .": quoted in Brock, *Justus von Liebig.*

95 . . . Lawes would contest Liebig's conclusions . . . : Johnston, "Liebig and the Rothamsted Experiments."

95 "entirely devoid of value . . ."; "it is all humbug . . .": quoted in Brock, *Justus von Liebig.*

95 "Now why on Earth did the Germans bomb Broadbalk?": Grey, *Rothamsted Experimental Station.*

96 . . . grassland plots at Cedar Creek in Minnesota: www.lter.umn.edu/.

96 "[t]here are no other long-term studies of this kind in existence . . .": D. Tilman, quoted in J. Pickrell, "Where the Grass Never Stops Growing," *Science* 293 (2001): 625.

96 . . . I found that each extra metric ton of hay . . . : J. W. Silvertown, "The Dynamics of a Grassland Ecosystem: Botanical Equilibrium in the Park Grass Experiment," *Journal of Applied Ecology* 17 (1980): 491–504.

97 . . . *Anthoxanthum odoratum* . . . evolved tolerance . . . : R. W. Snaydon, "Rapid Population Differentiation in a Mosaic Environment. I: The Response of *Anthoxanthum odoratum* to Soils," *Evolution* 24 (1970): 257–69.

98 It is estimated that, back in 1850, . . . : K. W. T. Goulding et al., "Nitrogen Deposition and Its Contribution to Nitrogen Cycling and Associated Soil Processes," *New Phytologist* 139 (1998): 49–58.

98 . . . she published in *Science* magazine her discovery . . . : C. J. Stevens et al., "Impact of Nitrogen Deposition on the Species Richness of Grasslands," *Science* 303 (2004): 1876–89.

99 In Dutch heathlands . . . : R. Bobbink et al., "The Effects of Air-Borne Nitrogen Pollutants on Species Diversity in Natural and Semi-Natural European Vegetation," *Journal of Ecology* 86 (1998): 717–38.

100 At Harvard Forest in Massachusetts . . . : A. H. Magill et al., "Ecosystem Response to 15 Years of Chronic Nitrogen Additions at the Harvard Forest LTER, Massachusetts, USA," *Forest Ecology and Management* 196 (2004): 7–28.

100 . . . red spruce (*Picea rubens*) is disappearing . . . : J. J. Battles et al., "Red Spruce Death: Effects on Forest Composition and Structure on Whiteface Mountain, New-York," *Bulletin of the Torrey Botanical Club* 119 (1992): 418–30; E. K. Miller et al., "Atmospheric Deposition to Forests along an Elevational Gradient at Whiteface-Mountain, NY, USA," *Atmospheric Environment* 27 (1993): 2121–36.

100 Experimental nitrogen additions to desert plots . . . : M. L. Brooks, "Effects

of Increased Soil Nitrogen on the Dominance of Alien Annual Plants in the Mojave Desert," *Journal of Applied Ecology* 40 (2003): 344–53.

101 . . . addition of nitrogen fertilizer to a stand of diffuse knapweed . . . : K. D. LeJeune and T. R. Seastedt, "*Centaurea* Species: The Forb That Won the West," *Conservation Biology* 15 (2001): 1568–74.

CHAPTER 8: FLORIDA!

General Sources

A species-by-species account with color photographs of each one of Florida's principal alien plants can be found in K. A. Langeland and K. C. Burks, eds., *Identification and Biology of Non-Native Plants in Florida's Natural Areas* (Gainesville: University of Florida Press, 1998). There are numerous accounts of invasive organisms and their environmental effects. The most pertinent to this chapter is D. Simberloff, D. C. Schmitz, and T. C. Brown, eds., *Strangers in Paradise: Impact and Management of Nonindigenous Species in Florida* (Washington, D.C.: Island Press, 1997). A very readable and quite frightening account of the situation in Australia is Tim Low's *Feral Future* (New York: Penguin Books, 1999).

Notes

102 "[T]here are the adventive plants . . .": quoted in Langeland and Burks, *Identification and Biology of Non-native Plants in Florida's Natural Areas.*

104 . . . a scientific monograph devoted to them . . . : G. E. Woolfenden and J. W. Fitzpatrick, *The Florida Scrub-Jay: Demography of a Co-operative Breeding Bird* (Princeton: Princeton University Press, 1985).

105 "For sixty miles or so south of Lake Okeechobee . . .": M. S. Douglas, *The Everglades: River of Grass* (Marietta, Ga.: Mockingbird Books, 1974).

106 "I drove past dark forests of paperbarks . . .": Low, *Feral Future.*

112 "We have the apparent double anomaly . . .": quoted in A. W. Crosby, *Ecological Imperialism* (Cambridge: Cambridge University Press, 1993).

114 In Europe, purple loosestrife is plagued . . . : D. Q. Thompson et al., *Spread, Impact, and Control of Purple Loosestrife* (Lythrum salicaria) *in North American Wetlands* (Washington, D.C.: U.S. Fish and Wildlife Service, 1987). Also available at Northern Prairie Wildlife Research Center's website: www.npwrc.usgs.gov/resource/1999/loosstrf/loosstrf.htm (Version 04JUN99).

115 "Does it not hurt your Yankee pride . . .": quoted in Crosby, *Ecological Imperialism.*

115 Imports of wool were a particularly rich source of seeds . . . : E. J. Salisbury, *Weeds and Aliens* (London: Collins, 1961).

115 . . . Ida Hayward and Claridge Druce published a book . . . : I. M. Hayward and G. C. Druce, *The Adventive Flora of Tweedside* (Arbroath: Buncle, 1919).

116 . . . on a larger scale throughout Europe: E. F. Weber, "The Alien Flora of Europe: a Taxonomic and Biogeographic Review," *Journal of Vegetation Science* 8 (1997): 565–72.

116 "prodigious number of plants in our gardens . . .": C. Darwin, *The Origin of*

Species by Means of Natural Selection, 1st ed. (London: John Murray, 1859), ch. 3.

116 . . . not one alien grassland species has managed to reciprocate: M. J. Crawley et al., "Comparative Ecology of the Native and Alien Floras of the British Isles," *Philosophical Transactions of the Royal Society of London Series B—Biological Sciences* 351 (1996): 1251–59.

CHAPTER 9: NEW DEMONS?

General Sources

For a very readable account of genetic modification (GM) issues in relation to potatoes, see M. Polan, *The Botany of Desire* (New York: Random House, 2001), ch. 4. An impartial website on GM issues run by the Pew Initiative on Food and Biotechnology may be found at www.pewagbiotech.org.

Notes

118 . . . a domesticate of the wild species *Brassica oleracea* . . . : D. Zohary and M. Hopf, *Domestication of Plants in the Old World*, 2d ed. (Oxford: Oxford University Press, 1994).

121 . . . willows and poplars are ancient polyploids: P. S. Soltis and D. E. Soltis, "The Role of Genetic and Genomic Attributes in the Success of Polyploids," *Proceedings of the National Academy of Sciences of the USA* 97 (2000): 7051–57.

122 . . . where it is encroaching on mudflats . . . : S. M. Percival et al., "Intertidal Habitat Loss and Wildfowl Numbers: Applications of a Spatial Depletion Model," *Journal of Applied Ecology* 35 (1998): 57–63.

122 Its American parent has been transported . . . : D. R. Ayres et al., "Extent and Degree of Hybridization between Exotic (*Spartina alterniflora*) and Native (*S. foliosa*) Cordgrass (Poaceae) in California, USA Determined by Random Amplified Polymorphic DNA (RAPDs)," *Molecular Ecology* 8 (1999): 1179–86.

122 . . . recent genetic analysis has vindicated them: L. H. Rieseberg et al., "Major Ecological Transitions in Wild Sunflowers Facilitated by Hybridization," *Science* 301 (2003): 1211–16.

122 The genes of *H. paradoxus* . . . : L. H. Rieseberg, "Hybrid Origins of Plant Species," *Annual Review of Ecology and Systematics* 28 (1997): 359–89.

123 . . . successes seem to be a comparative rarity: C. A. Stace, *Hybridization Flora of the British Isles* (Cambridge: Cambridge University Press, 1975). Also see N. C. Ellstrand et al., "Distribution of Spontaneous Plant Hybrids," *Proceedings of the National Academy of Sciences of the USA* 93 (1996): 5090–93.

123 . . . new species discovered by Joseph Pitton de Tournefort . . . : M. Campbell-Culver, *The Origin of Plants* (London: Hodder, 2001).

123 . . . Victorian gardeners who liberally planted *R. ponticum* . . . : M. Walters, *Wild and Garden Plants* (London: Harper Collins, 1993).

124 . . . *R. ponticum* evidently hybridized with *R. catawbiense* . . . : R. I. Milne and R. J. Abbott, "Origin and Evolution of Invasive Naturalized Material of

Rhododendron ponticum L. in the British Isles," *Molecular Ecology* 9 (2000): 541–56.

124 Two species of *Casuarina* . . . : T. Low, *Feral Future* (New York: Penguin Books, 1999).

125 "It's a myth to think . . .": in *Biotech Branches Out*, a summary of a meeting organized by the Pew Trusts and available at www.pewagbiotech.org/events/1204/branch-summary.pdf.

127 . . . GM canola were being grown this way in western Canada . . . : S. C. H. Barrett et al., "Special Issue by an Expert Panel of the Royal Society of Canada—Elements of Precaution: Recommendations for the Regulation of Food Biotechnology in Canada," *Journal of Toxicology and Environmental Health—Part A* 64 (2001): issues 1–2.

128 Experiments in Britain in which conventional and GM varieties . . . : M. J. Crawley et al., "Transgenic Crops in Natural Habitats," *Nature* 409 (2001): 682–83.

129 . . . escape from crops into wild relatives . . . : N. C. Ellstrand, *Dangerous Liasons? When Cultivated Plants Mate with Their Wild Relatives* (Baltimore: Johns Hopkins University Press, 2003).

129 Experimental transfer of *Bt* from crop sunflowers . . . : A. A. Snow et al., "A *Bt* Transgene Reduces Herbivory and Enhances Fecundity in Wild Sunflowers," *Ecological Applications* 13 (2003): 279–86.

129 . . . harmful effects of *Bt* corn pollen on monarchs were negligible . . . : M. K. Sears et al., "Impact of *Bt* Corn Pollen on Monarch Butterfly Populations: A Risk Assessment," *Proceedings of the National Academy of Sciences of the USA* 98 (2001): 11937–42.

130 There also appear to be health benefits . . . : J. Huang et al., "Plant Biotechnology in China," *Science* 295 (2002): 674–77.

130 The studies were at the farm scale . . . : D. B. Roy et al., "Invertebrates and Vegetation of Field Margins Adjacent to Crops Subject Contrasting Herbicide Regimes in the Farm Scale Evaluations of Genetically Modified Herbicide-Tolerant Crops," *Proceedings of the Royal Society of London* 358 (2003): 1879–98.

130 . . . question as to whether the beneficial effects . . . would be long-lasting . . . : R. P. Freckleton et al., "Deciding the Future of GM Crops in Europe," *Science* 302 (2003): 994–96.

131 Trees genetically modified to resist the elm disease . . . : C. C. Mann and M. L. Plummer, "Forest Biotech Edges Out of the Lab," *Science* 295 (2002): 1626–28.

CHAPTER 10: THE END OF EDEN

General Sources

Environmental books are ten-a-penny, but two stand out from the crowd: David Quammen's *The Song of the Dodo* (London: Hutchinson, 1996), and *The Future of Life* (Boston: Little, Brown, 2002), by Edward O. Wilson.

Notes

133 "We use the wood of this tree for tables": All statements by Lorenzo Martinez and Felix Castellaños are from the transcript of television program S326/15, *Food from the Rainforest*. Open University copyright. Filmed in 1986.

135 . . . forest in Los Tuxtlas was being lost . . .": R. Dirzo and M. C. Garcia, "Rates of Deforestation in Los Tuxtlas, a Neotropical Area in Southeast Mexico," *Conservation Biology* 6 (1992): 84–90.

135 Half of these plants are found nowhere else . . . : R. Dirzo, *Mexican Diversity of Flora*, 2d ed. (Mexico City: Cemex, 1996).

136 . . . Rodolfo Dirzo suggests . . . : E. Mendoza and R. Dirzo, "Deforestation in Lacandonia (Southeast Mexico): Evidence for the Declaration of the Northernmost Tropical Hot-Spot," *Biodiversity and Conservation* 8 (1999): 1621–41.

136 "is now somewhat disguised by intrusive strips . . .": Thomas Hardy in the preface to *Return of the Native* (London: Macmillan, 1895).

136 . . . reaching almost Dalmatian spottiness and scarcity by 1978: N. Webb, *Heathlands* (London: Collins, 1986).

136 . . . Dorset heathland habitat is being extensively restored . . . : www.english-nature.org.uk/thh/default (accessed 19 September 2002).

136–137 . . . Peter Marren trawled through published floras . . . : P. Marren, "What Time Hath Stole Away: Local Extinctions in Our Native Flora," *British Wildlife* (June 2001): 305–10.

138 WCMC collates information . . . : B. Groombridge and M. D. Jenkins, *Global Biodiversity: Earth's Living Resources in the 21st Century* (Cambridge: World Conservation Press, 2000).

138 . . . the four hundred thousand plants known: R. Govaerts, "How Many Species of Seed Plants Are There? *Taxon* 50 (2001): 1085–90.

139 . . . a population of only forty-three adult individuals: A. Beattie and P. R. Ehrlich, *Wild Solutions* (New Haven: Yale University Press, 2001). Population size as of 13 September 2002 (www.rbgsyd.gov.au/html/Wollemi/Facts.html).

140 The renovated plumbing of the Florida Everglades ecosystem . . . : World Resources Institute, *World Resources 2000–2001: People and Ecosystems: The Fraying Web of Life* (Washington, D.C.: World Resources Institute, 2000).

142 "The problem is, I could sit here . . .": quoted in W. Allen, *Green Phoenix: Restoring the Tropical Forests of Guanacaste, Costa Rica* (Oxford: Oxford University Press, 2001).

142 "Set aside your random research . . .": D. H. Janzen, "The Future of Tropical Ecology," *Annual Review of Ecology and Systematics* 17 (1986): 305–24.

144 "Choose an appropriate site . . .": D. H. Janzen, "Tropical Ecological and Biocultural Restoration," *Science* 239 (1988): 243–44.

144 "Restoring complex tropical wildlands is primarily a social endeavour . . .": D. H. Janzen, "How to Grow a Wildland: The Gardenification of Nature," in *Nature and Human Society*, ed. P. H. Raven, 521–29 (Washington, D.C.: National Academy Press, 2000).

144 . . . William Allen in his book *Green Phoenix* . . . : Allen, *Green Phoenix*.

Index